The Mendelian Revolution

**Also by Peter J. Bowler
from Johns Hopkins**

The Eclipse of Darwinism: Anti-Darwinian Evolution Theories in the Decades around 1900 (1983)

Theories of Human Evolution: A Century of Debate, 1844–1944 (1986)

The Non-Darwinian Revolution: Reinterpreting a Historical Myth (1988)

The Mendelian Revolution

THE EMERGENCE OF HEREDITARIAN CONCEPTS
IN MODERN SCIENCE AND SOCIETY

Peter J. Bowler

The Johns Hopkins University Press
Baltimore

Printed in Great Britain

First published in 1989

The Johns Hopkins University Press, 701 West 40th Street, Baltimore, Maryland 21211

Library of Congress Cataloging-in-Publication Data

Bowler, Peter J.
    The Mendelian revolution: the emergence of hereditarian concepts in modern science and society/Peter J. Bowler.
       p.   cm.
    Bibliography: p.
    Includes index.
    ISBN 0–8018–3888–6
    1. Genetics—History.   2. Biology—History.
    3. Mendel, Gregor, 1822–1884.   I. Title.
    QH428.B69   1989   89–30914
    575.1′09—dc19   CIP

Contents

Preface

This book is neither a biography of Mendel nor a history of genetics in the conventional sense. It is an account of the major changes in our thinking about heredity and reproduction which have taken place over the last couple of centuries, with emphasis on the emergence of genetics as a new scientific discipline in the early twentieth century. As the first chapter explains, historians of science are engaged in a dramatic reformulation of their thinking on this episode in the development of modern biology. This book attempts to summarize the results of this historiographical revolution for the non-specialist. Its purpose is to allow readers with little or no knowledge of biological technicalities to understand the development of modern concepts of heredity and their implications.

Since I myself am a non-specialist in this area, the book could not have been written without the help and encouragement of a host of scholars engaged in specialized research. My intellectual debts are recorded in the references to numerous items in the bibliography, but I am especially grateful to those who have provided advice, offprints and preprints of unpublished research: Garland Allen, Richard Burian, Len Callender, Jon Harwood, Jon Hodge, Robert Olby and Jane Maienschein. Robert Olby and Jon Hodge read the original manuscript and responded with valuable comments, although I am responsible for any remaining errors and all idiosyncracies of interpretation. If I can arouse some wider interest in this area of the history of biology, I shall have partially repaid my debt to the specialists whose work I have drawn upon so freely.

It will probably be obvious from the book's structure that my own field of interest is the development of evolution theory, an area which intersects with but does not overlap the history of genetics. As a partial outsider, I have tried to present the material in a way that makes sense to me – and may even prove illuminating to some specialists. If there is anything original in the following pages, it arises from my efforts to show that the advent of Mendelism can be seen as a conceptual revolution whose implications are at least as fundamental as those of the so-called 'Darwinian revolution'. My own work on evolutionism has convinced me that some of the consequences usually attributed to the impact of Darwin's theory should in fact be explained in terms of later events associated with the rise of genetics. In general, I have found that the reinterpretations being offered by historians of genetics tie in neatly with my own claims about the development of evolutionism. When The Athlone Press invited me to write a non-specialist survey of current thinking about the origins of Mendelism, I was thus presented with an offer I could hardly refuse.

1

Mendelism: Discovery or Invention?

This book provides a history of the early phases in the development of modern genetics. It will cover the 'rediscovery' of Gregor Mendel's laws of heredity in 1900 and the emergence of a new science of inheritance, at first often called Mendelism, but increasingly known as genetics. It will explore the links established by T. H. Morgan and others between Mendel's laws and the behaviour of the chromosomes in the cell nucleus, links which created the modern concept of the gene as a material unit responsible for encoding the biological information transmitted from parent to offspring. The resulting 'classical genetics' established itself as a triumph of modern biology, apparently vindicating the scientists' claim that their method can unlock the secrets of nature and give mankind control of the material world.

Classical genetics established the foundations upon which modern molecular biology has been built. Once the structure and function of DNA was deciphered, scientists could tackle the problem of *how* the genetic information was coded and used to construct the body of a new living organism. In the era of biotechnology, we take it for granted that the scientists can manufacture organisms with whatever characters we require. This book does not cover the emergence of molecular biology, but the importance of classical genetics in establishing modern attitudes toward biology and its social consequences cannot be underestimated. Mendelism was a stepping stone towards the exciting – and perhaps frightening – world of late twentieth-century biology. If we wish to understand the role played by science in the complex world we live in, Mendelism's origins will certainly repay serious investigation.

Historians of science have not been idle in this respect. The history of genetics cannot rival the 'Darwin industry' in the number of academic publications it generates, but it has become a very active area of research. Much of the existing literature, however, is aimed at an academic readership, and its conclusions do not readily percolate through to the world of the nonspecialist. This book is intended to translate the findings of modern historians into a language accessible to the reader with little or no scientific training. It is a history of genetics conceived on a broad scale, concentrating on the major conceptual and social developments, not on technical details that may be of interest only to biologists. And – perhaps to the dismay of the biologists themselves – it will show how the conventional image of genetics as a triumph of scientific discovery needs to be qualified if not completely abandoned in the light of historical research.

Challenging the Myth

Scientists in general tend to have little real interest in the history of their discipline, but most of them accept an outline of the great discoveries that paved the way towards modern knowledge. Historical sketches embodying what might be called the 'orthodox' version of a science's origins are to be found at the beginning of many textbooks. In the case of genetics we have several books outlining the development of the field in more detail, written by geneticists with an enhanced need to provide their science with a historical foundation (Carlson, 1966; Dunn, 1965; Stubbe, 1972; Sturtevant, 1965). The coincidence in dates (the German edition of Stubbe's book appeared in 1965) suggests that these were all written, perhaps unconsciously, to celebrate approximately fifty years of genetics's existence as an independent branch of biology. The orthodox image of the history of genetics outlined by these books can be summarized as follows.

The study of heredity remained in a state of confusion throughout the pre-Mendelian era. Myths and old wives' tales served only to obscure a subject which was, by its very nature, extremely complex. Strange ideas were able to flourish, perhaps the strangest of all being the 'preformation theory' of the period around 1700, according to which the embryo grew from a perfectly formed miniature already present in the mother's ovum (or the father's sperm). The nineteenth century abandoned such bizarre notions, but adopted a confused theory of 'blending inheritance' in which the offspring's characters were always intermediate between those of its parents. Most authorities still thought that heredity was 'soft', i.e. that the transmission of characters to the offspring could be modified by changes taking place in the parents' bodies due to new habits or a new environment. Even Charles Darwin accepted these incorrect ideas and enshrined them in his own theory of 'pangenesis'. In so doing Darwin not only helped to ensure that Mendel's pioneering work was obscured, but also created a major problem for his broader theory of evolution by natural selection.

Only in the late nineteenth century did things begin to change for the better. Improved knowledge of the cellular structure of living tissue helped to focus attention onto the chromosomes, minute rod-like structures contained within the cell nucleus, as the bearers of heredity. August Weismann advanced the notion of 'hard' heredity – the modern view that characters are transmitted unchanged from one generation to the next and cannot be influenced by changes in the parent's body. The body acts merely as a 'host' for genetic material that will be transmitted to future generations. These new developments at last created a climate of opinion in which the work of the Austrian monk, Gregor Mendel, could be appreciated. Mendel's classic experiments with peas had demonstrated the laws of inheritance which now bear his name, thereby undermining the plausibility of blending heredity and establishing the existence of discrete genetic characters. Published in 1865, these laws had been ignored, partly because Mendel had written in an obscure local journal and partly because the excitement generated by Darwin's theory had kept attention focused on outdated models.

By the turn of the century, the time was at last ripe for general recognition of the true nature of heredity as summed up in Mendel's laws. Thus it was in 1900 that three biologists, Carl Correns, Hugo De Vries and E. von Tschermak, independently rediscovered Mendel's laws and came to acknowledge the work of the now-dead pioneer. Although at first resisted by biologists unable to throw off outdated notions, the new approach gained an increasing number of converts and soon established itself as the basis for a new science of heredity. Soon the work of T. H. Morgan and his school, using the fruit fly, *Drosophila*, showed that the behaviour of the chromosomes during reproduction corresponded exactly to the transmission of characters according to Mendel's laws. Classical genetics could now exploit the concept of the gene as a discrete material unit on the chromosome, coding for a particular character that could be transmitted from parent to offspring, isolated from all external influence. The new theory resolved the difficulties inherent in Darwin's original formulation of natural selection, allowing the 'genetical theory of natural selection' to become the dominant paradigm in evolutionary biology.

Genetics ushered in a new era in the science of animal and plant breeding, and in the identification of hereditary defects in humans. Once the complexities of the transmission process were understood, the way was clear for the emergence of molecular biology, a second generation science that would explain the biochemical and physiological processes by which the hereditary information was coded and then unfolded into living tissue. Recognition of the 'double helix' structure of DNA (the chemical of which the genetic material is composed) by James Watson and Francis Crick in 1953 opened the way to yet greater understanding and control of the process of inheritance (Watson, 1968; Olby, 1974).

The key element in this orthodox story is the notion of discovery. The account is based on the assumption that genetics – like all sciences – reveals true information about the real world. The scientist discovers the truth by using a method of investigation that lifts his enterprise above the subjective world of human affairs, with its endless cultural, social and ideological debates. The study of heredity was originally confused because early biologists were unable or unwilling to use the scientific method to challenge the existing myths about how inheritance worked. Mendel showed how the problem could be tackled; he discovered the true laws of heredity, only to find that the cultural environment still prevented appreciation of his work. Only when a battery of indirect scientific assaults had destroyed the credibility of the old notions did the rediscovery of Mendel's laws become possible. Once that point was reached, however, three scientists independently hit upon the truth within a short period of time. Their work ensured that Mendelism became a permanent component within the arsenal of human knowledge, the firm foundation for a series of later discoveries. At each step in the process, the discovery of new facts allowed an increasing level of control over nature. As with all sciences, the success of our technology demonstrates the truth of the information discovered.

Several aspects of the orthodox history of genetics help to reinforce the image of scientific discovery as the antidote to ignorance and superstition. Many sciences have a founder who is accorded almost heroic status as the first person to introduce the light of rational investigation into the field. In Mendel's case, the heroic image is strengthened rather than obscured by his contemporaries' unwillingness to accept his demonstration of what we now know to be the true laws of inheritance. Mendel lost the battle against ignorance, at least temporarily, but the sense of tragedy only highlights his achievement. The multiple rediscovery of his laws shows that once the cultural obstacles have been overcome, the path to the truth is open to all, and the scientific method produces instant results. The dovetailing of Mendelian breeding experiments with research on the cell nucleus confirms that different areas of biology can lead independently towards the same conclusions, which must therefore be accepted as incontrovertible truths. In the modern world, the often repeated claim that the structure of DNA represents the 'secret of life' helps to drive home the claim that genetics has, in the end, brought us face to face with the most basic of nature's operations.

The majority of scientists argue that science is a disinterested search for the truth, just as the majority of geneticists accept that their discipline is based on a series of factual discoveries. Until recently, most laypersons were inclined to accept both this general interpretation of the scientific method and the specific image of genetics as a good example of how the method works to the benefit of all. In recent years, however, the independence and objectivity of science has increasingly been questioned even by those who have no direct insights into its operations. Everyone can now see that science is a political activity: controversies over nuclear weapons, nuclear power and the state of the environment have shown that scientists do not speak with one voice. Far from presenting a simple vision of the truth, scientists seem to speak out on both sides of a disputed question, with the side they choose having a suspicious tendency to be linked with the source of their salaries and research grants. Nor is genetics free from this tendency. We are all aware of debates over the safety of genetic engineering and over the degree to which human mental capacities are determined by inheritance. Collectively, these debates raise doubts about the objectivity of science. More specifically, the disputed applications of genetics must lead us to wonder if the history of genetics, conceived as a series of factual discoveries, is an adequate model of what actually happened.

The purpose of this book is to suggest that current research in the history and sociology of science confirms the need to rethink the orthodox view of science in general and the origins of genetics in particular. The challenge can be presented at three different levels, all contributing to a new interpretation of the way science operates. The three levels are:

1. *Conceptual*   No philosopher of science now accepts the old idea that the scientific method operates through the simple accumulation of facts about nature. Facts only appear as facts within an appropriate conceptual scheme: what is fundamentally important when viewed from one perspective may seem trivially irrelevant from another. Mendel's laws appear as important discoveries only within a theoretical model of heredity in which *(a)* the transmission of characters from one generation to the next constitutes a distinct and worthwhile field of study and *(b)* it is assumed that the characters can be treated as distinct units. History reveals that neither of these conditions was satisfied in the pre-Mendelian era. Darwin and most of his contemporaries accepted a 'developmental' model of heredity which made no distinction between the transmission of characters from one generation to the next and the process by which the characters are produced in the growing organism. Theirs was a theory of 'generation' (Hodge, 1985): it treated reproduction and growth as parts of an integrated biological process and formed a conceptual scheme in which a separate study of transmission was simply inconceivable. The scheme also made it impossible to think in terms of discrete units of inheritance, or to believe that characters were transmitted unchanged from one generation to the next.

Acceptance of Mendel's laws thus depended not on the discovery of facts, but on the creation of a new conceptual scheme within which laws of that kind could make sense. To dismiss pre-Mendelian theory as mere ignorance and superstition obscures the fact that major thinkers such as Darwin were able to use it as a vehicle for making important theoretical innovations in other areas of biology. Darwin's mechanism of natural selection was *not* rendered impotent by its lack of a genetic model of heredity. It functioned perfectly well within the pre-Mendelian scheme (although it encountered other problems that had nothing to do with blending inheritance). The advent of Mendelism represented not the simple discovery of certain factual relationships, but a conceptual revolution of major proportions. Genetics was the product of new ideas not only about heredity itself, but also about the relationship between heredity and other biological phenomena.

One purpose of this book is to emphasize the sheer scope of what might be called the 'Mendelian revolution' in biology. We are all familiar with the notion of a Darwinian revolution brought about by recognition of the radical implications of evolution theory. Here was a 'discovery' with the power to reshape conventional thought on a host of philosophical and even theological issues. Yet Darwin left certain aspects of the traditional view of life intact, and the effects of his revolution were thus far more restricted than has often been supposed (Bowler, 1988). Evolutionism flourished in the late nineteenth century, but it was often conceived within a theoretical framework that allowed the retention of teleological values that Darwinism is popularly supposed to have destroyed. It was precisely the developmental view of reproduction and heredity that allowed many biologists (and non-scientists) to continue believing that evolution was a purposeful process. Darwin tried and ultimately failed to dislodge the idea that evolution progressed inevitably towards a morally significant goal. The fact that modern biologists do not see evolution as directed towards predetermined goals is largely the result of the conceptual revolution associated with the advent of Mendelism. Although its effects are not as immediately visible as those of the Darwinian debate, the Mendelian revolution may ultimately have to be regarded as at least as important a transformation in our ideas about life (including, of course, human life). The claim that such a revolution could be accomplished simply by making an experimental discovery is, on the face of it, absurd.

If historians are increasingly recognizing the importance of the early twentieth-century revolution in heredity theory, what have they got to say about Mendel's 'premature' discovery? On the traditional view that genetics depends upon a simple recognition of facts, the refusal of Mendel's contemporaries to see his work as the basis for a new science is something of a puzzle. But once we accept that genetics emerged as part of a conceptual revolution, it becomes obvious that Mendel's fellow biologists would not have been able to recognize implications that would only become apparent over thirty years later. Indeed, on the view that science advances only through theoretical revolutions, the real puzzle is how Mendel could have developed concepts that were so far ahead of their time. Some historians of genetics now suppose that Mendel was not, in fact, attempting to pioneer a new theory of heredity. His real concern was the hybridization of species as an alternative to evolution, and his discovery of regularities in the inheritance of characters was merely a by-product of a research programme that would not have made sense to his 'rediscoverers'. In the words of one historian of genetics, Mendel may not have been a Mendelian (Olby, 1979, 1985).

2. *Professional* One implication of the above analysis is that science must be seen as a social activity, not as a mechanical process for generating facts. At certain points in the history of science, someone has to generate a new conceptual scheme and then persuade others that it makes more sense than the old one. The promulgation of a new theory is thus inevitably a process requiring social interaction within the scientific community. It might be argued that success in this enterprise will depend solely upon the number of experimental facts at the disposal of the theory's supporters. But the historical evidence suggests that in the hurly-burly surrounding most theoretical debates, what is accepted as a relevant fact may depend upon the argumentative skills of the various protagonists. New theories seldom offer a perfect solution to current problems, hence their acceptance or rejection may, in the end, turn upon their supporters' abilities in the field of public relations (Hull, 1978). Good PR work may not be able to sell a bad theory, but a potentially good one may find its acceptance blocked if its proponents cannot play the game of scientific politics. They must adopt a workable strategy for converting others, undermining the influence of opponents, gaining access to journals and research grants, and all the other activities required to ensure an expanding role within the scientific community.

The emergence of genetics was a particularly complex process because it required not a revolution within an existing science, but the creation of an entirely new discipline. The study of heredity as an isolated phenomenon had not been accepted as a viable scientific activity in the pre-Mendelian era. The early geneticists had to establish a place for themselves: they needed university positions, journals, government and industrial funding – all the paraphernalia of modern 'big science'. We know that they were successful, as a matter of fact, and this may encourage the view that the success was an inevitable product of genetics's ability to plug a gap in the range of biological disciplines. But the alternative view that genetics is *not* an intrinsically 'obvious' field of study is supported by the fact that its establishment varied from country to country. Science is often assumed to be a completely international activity, but it does have certain national 'flavours' – and in some cases the differences amount to major variations in what is recognized as an important contribution to knowledge. Classical genetics achieved its greatest institutional success in America; it was rather less clearly defined in Britain, much less so in Germany, and almost totally nonexistent in France. Genetics, as it came to exist in the English-speaking world, was the product of social developments within the scientific community that were not necessarily copied by other countries. Presumably scientists in those other countries were not aware of any universal consensus that the new approach to heredity constituted a viable field of study.

These facts confirm that science is an essentially social activity, and suggest that we should look out for social constraints on the kind of theorizing permitted within a new discipline. In fact there was a 'struggle for authority' within the ranks of the early geneticists, by which certain potentially interesting theoretical avenues were branded as heretical and deliberately excluded from scientific discourse (Sapp, 1987). The possibility that some inheritance could be transmitted by the cytoplasm (the material surrounding the cell nucleus) was dismissed as a vestige of outdated speculations. The growth of science does not represent an open-ended assault on the unknown: it is shaped by the theoretical models considered acceptable by the community that defines the field of study. In this respect, at least, professional scientists resemble the disciples of a new religion.

3. *Ideological*   The fact that science advances through conceptual revolutions brought about within the scientific community should alert us to the possibility that new theories may interact with social developments on a wider scale. No one doubts that Darwin's theory of evolution had a major impact on Victorian society through its challenge to the notion of divine creation and its promotion of new

moral values which are often called 'social Darwinism'. But these consequences were no mere spin-off from a factual discovery. Darwinism in science was made possible by the emergence of new social values (Young, 1985). It is often claimed that the idea of natural selection represents a projection onto nature of the competitive value system of Victorian capitalism. Historians disagree over the strength of the analogy, but few would now deny that Darwin drew some inspiration from the ideology of competitive individualism known as *laissez-faire* economics.

On this model, new laws or theories are not simply 'discovered' (implying a real world to which we have direct access via the scientific method). They are *invented* to satisfy the cultural values of the scientists and of the public with whom they must interact. The new conceptual scheme is not determined solely by the facts available – although it must, of course, be able to accommodate those facts if it is to serve a useful function in science. But the conceptual structure of the theory may have been designed (consciously or unconsciously) to shape our perception of the facts in accordance with the subjective values of the scientists whose imagination is the ultimate source of theoretical innovation.

If the advent of Mendelism represents – as claimed above – a revolution in our views about the nature of life as fundamental as that proposed by Darwin, then it would be surprising if it too were not influenced by ideological factors. In fact there is an obvious candidate for the source of such an interaction with society. The rediscovery of Mendel's laws coincided with the emergence onto the political scene of what may be termed 'hereditarian' social policies. At the turn of the century, politicians and social commentators from certain backgrounds began to argue that human abilities are rigidly determined by heredity. Reform and the improvement of people's conditions could not produce any significant effect because the proportion of good and bad characteristics in the population is determined solely by inheritance. Through the eugenics movement, the advocates of this view of human nature began to argue the necessity for selective breeding policies designed to prevent those individuals with harmful characters from passing them on to future generations. As in the case of Darwinism, the link to science is extremely complex, but it is difficult to escape the feeling that the success of genetics came about at least in part because it could be presented as a plausible scientific foundation for these social attitudes.

The significance of the conceptual and ideological developments associated with the introduction of hereditarian ideas has often been underestimated. This has occurred partly because those who oppose the application of such ideas to human affairs seem anxious to trace their enemy as far back in the history of biology as possible. All too often the modern versions of hereditarianism are dismissed as merely revived 'social Darwinism' – a move which carries the obvious implication that Darwinism, not Mendelism, is the real source of such values. Yet the classic social Darwinism of the late nineteenth century was really an offshoot of Herbert Spencer's evolutionary philosophy, and Spencer was not a wholehearted hereditarian. True, he believed that the intellectual and moral capacities of the various human races were fixed by their biological constitutions – but so did almost everyone else in the nineteenth century, including the creationists. Within modern European society, however, Spencer did not hold that the individual's level of attainment is limited by inheritance. When he advocated un-restrained competition, he knew that a few unfortunates might be eliminated by starvation, but his real purpose was to ensure that everyone would be stimulated to try harder in order to escape the painful consequences of failure. The hereditarian ideas of the early twentieth century represented an ideological transformation be-cause they centred on the claim that one could *not* stimulate individuals to greater efforts. Each person's level of attainment was fixed by his or her biological character, as determined by heredity. The eugenics movement advocated state control of human fertility because it was feared that the biologically unfit were breeding like rabbits in the slums, in direct violation of Spencer's predictions.

The social consequences of hereditarianism have been less visible than the effects of the Darwinian revolution because the new interpretation of human nature was challenged from the start by opponents determined to show that human nature can be improved by reformed social conditions (Cravens, 1978). The modern social sciences emerged at the same time as genetics but were dedicated to the view that culture, not biological inheritance, determines how a person behaves. In the great debate over whether nature or nurture determines human character, the social sciences have upheld the importance of nurture. Reform is possible because a better upbringing creates a better person. Sociologists dismiss the attempt to show that character is determined by nature (i.e. by biological inheritance) as a conservative ploy designed to allow the rich to retain the illusion that their pre-eminence is guaranteed by superior ability. The controversies over sociobiology and over the claim that IQ is genetically determined show that such issues are still with us today. Is genetic determinism a device used by biologists to support the claim that they, and not the sociologists, should be allowed to govern our lives? Or do the social sciences deny the role of heredity merely to increase their own influence? Whichever side one takes in this argument, the role played by theories of inheritance is of crucial importance. By recognizing that Mendelism emerged in the period at which hereditarian ideology began a serious bid for influence, we accept the need to evaluate the new biological theory at least partly in terms of its social consequences.

To sum up the three levels of reinterpretation offered above, we have suggested that the orthodox history of genetics needs to be revised in a way that will allow the science to appear as a human activity, not a mechanical fact-gathering exercise. Whatever their experimental basis, the great 'discoveries' could not have been made without conceptual innovations that have changed our basic ideas about the nature of life. These new ideas had to be promoted within the scientific community, competing – not always successfully – with rivals for the money and influence needed to found a new discipline. The ideas themselves may have been shaped to some extent by the need to create a professional identity, and by changing values within society at large. Public interest in genetics was certainly influenced by the belief that it offered information of practical consequence for human affairs. The claim that the history of genetics can be seen as a series of factual discoveries is itself part of the geneticists' bid for power and influence, since it is designed to reinforce the belief that the science offers a source of indisputable facts about human nature. Without denying the important factual consequences that have flowed from the development of genetics, the history of the field will show that the new science was invented to serve human purposes – it did not grow automatically as a consequence of factual observations.

The New View of Science

The attitudes expressed in this reconstructed history of genetics will shock many biologists – and many other scientists who fear that their disciplines may be threatened by radical historians and sociologists. The orthodox history of the subject has not only been challenged, it has been held up as an example of the value system that supports the scientific establishment. Yet however radical this proposed reinterpretation might seem, it is but one illustration of the increasingly critical (or increasingly objective?) view of science that has been adopted by scholars who now take the scientific enterprise itself as their field of study. Assaults on the notion of scientific objectivity have been launched from a variety of backgrounds, and it may be worth while to show how the revaluation proposed in this book fits into the broader reinterpretation of science that has emerged in the last few decades.

In the new model, the most important aspect of science is not the discovery of facts but the creation of new theories. Obviously these theories must be related to the facts, and must aid in the discovery of new facts, but this does not mean that they represent a true model of reality that will never be displaced. Theories are only models or pictures of reality; some are better than others, but none carries a guarantee that is an exact copy. The implication is that the theory we now accept is *not* the only model capable of dealing with the real world. Nature is a system so complex that no creation of the human mind will ever plumb its depths completely. We have only approximations, and in any particular area there will be more than one way of approximating to reality, just as there is more than one way of mapping an area of land. Other models could cope just as well as the one we now work with, and scientific progress occurs precisely because the inadequacy of existing models occasionally becomes so obvious that new approaches must be sought out. The possibility that the choice of models may be influenced by nonscientific factors is a necessary by-product of the view that theories are invented rather than discovered.

The impossibility of depicting the scientific method as simple fact-gathering has long been recognized even by philosophers of science who would be horrified by the claim that external factors influenced the choice of theories. Nature is simply too complex for anyone to see regularities or laws simply emerging from the welter of information that bombards our senses. The scientist must think ahead of observation. He or she must define a problem and then try to guess how nature *might* behave in this particular area. Only then are experiment and observation brought in to test whether or not the guess is along the right lines. This is the 'hypothetico-deductive' method (Hempel, 1966). The scientist first sets up a hypothesis and then deduces consequences from that hypothesis that are empirically testable (i.e. testable by experiment and observation). If the tests disprove the hypothesis, it is of course discarded and a new one sought. If the experiments are successful, the hypothesis can be accepted as a provisional working model suitable for further tests in new areas. Laws and theories are hypothetical generalizations that have so far passed all or most of the tests to which they have been subjected. The hypothetico-deductive method implies, however, that the law or theory should *never* be accepted as 'true'. There is always a chance that the next test will be unsuccessful.

The claim that science provides only provisional knowledge of the external world has been enshrined in the philosophy of Sir Karl Popper (1959). Popper's greatest concern is to demarcate between science and pseudo-science (subjects such as astrology which adopt a spurious air of scientific authority). The proposal and testing of hypotheses is still central to this philosophy, but Popper emphasizes that the chief characteristic of a truly scientific hypothesis is its testability or 'falsifiability'. All proposals must be capable of being tested in a way that will immediately demonstrate any inadequacy. The presence of 'fudge factors' that would allow *any* experimental result to be reconciled with the hypothesis is a sure sign that the discipline concerned is a pseudo-science. Despite his reinforcement of the view that science can never get at the ultimate truth about nature, the scientists themselves have adopted Popper's philosophy because it distinguishes their endeavour from all other forms of human activity. Science alone dares to accept that its conclusions must be rejected when necessary, and is thus the only objective way of approximating to knowledge of the real world.

Popper accepts that all hypotheses, however successful in the past, will ultimately be falsified and replaced with something better. Indeed, he implies that scientists normally strive to falsify rather than confirm their own inventions. This is a view which most historians – aware of the bitter debates that have surrounded the introduction of new ideas – find hard to accept. The first major challenge to Popper's vision of scientific objectivity came in Thomas S. Kuhn's analysis of scientific revolutions (1962). Kuhn sees science as a social phenomenon governed by codes of behaviour similar to those we expect to find in other human activities. For most of the time, scientists are doing 'normal science' in which the theoretical framework or 'paradigm' is taken for granted. Far from trying to falsify the paradigm, the scientists have every confidence in its validity and cheerfully claim that it represents a true picture of nature. Their experiments and observations are designed only to extend the range of phenomena to which the paradigm can be applied. The paradigm defines not only what will be considered as acceptable answers to experimental questions, but which questions are regarded as meaningful. The process of scientific education is intended to indoctrinate students into ready compliance with the paradigm, and anyone who violates the accepted standards is dismissed as a crank, someone who is by definition outside the scientific community.

Kuhn's book became a focus for debate because it implied that at times scientists turn a blind eye to phenomena that actually falsify the existing paradigm. Anomalies are swept under the carpet and ignored in favour of other areas where the paradigm is still a successful guide to research. Yet Kuhn too accepts a kind of objectivity, since his concept of scientific revolution was based on the belief that anomalies will eventually become too numerous to ignore. Science then enters a state of crisis which is resolved only by the appearance of a new hypothesis that will deal successfully with the phenomena that were incompatible with the old theory. Once such a new theory has emerged, younger scientists convert to it quite readily, although older ones may lack the intellectual flexibility and are left to die off unconverted. Eventually the theory becomes the new paradigm and begins its own period of domination over scientific thinking.

Historians and sociologists of science did not agree with the whole of Kuhn's analysis, but in general they welcomed his emphasis on science as an activity conducted by human beings who often behave in an apparently irrational manner (by Popper's standards) dictated by the values of the group to which they belong. The fact that paradigms are entities with definite beginnings and ends turns the history of science into a genuinely *historical* discipline, since it implies that one can only understand the science of a past era by trying to think oneself into the conceptual scheme of the then-dominant paradigm. Trying to evaluate the past by modern standards – depicting the past as a series of discoveries each representing a step towards our modern level of knowledge – can only lead to misunderstanding. The great discoveries of the past were not necessarily made by scientists who thought exactly as we do today, and their work has had to be reinterpreted by later generations in terms of subsequent paradigms. As a branch of the history of ideas, the history of science now routinely takes it for granted that the orthodox stories of past achievements contained in science textbooks are unreliable, since they inevitably distort the past in order to conceal the role played by earlier paradigms.

Apart from recognizing that the science of past eras may have been built on different conceptual foundations, historians have also turned their attention to the social structure of the scientific community. Their work thus interacts with that of the sociologists of science, who take as their field of study the functioning of the modern scientific community (for surveys see Barnes and Edge, 1982; Knorr-Cetina and Mulkay, 1983; Mulkay, 1979). The concept of a global scientific community is really an unworkable abstraction: scientists actually function within much smaller groups or networks, each of which accepts certain preconceived ideas about methodology and theoretical approach. These networks define what it is to be a geneticist, a molecular biologist, or whatever. Individual scientists or research groups within the network compete to see which can influence the direction of research to their own advantage. Investigation of modern scientific disciplines has revealed how closely knit they can be, and how easy it is for personal influence to control what does or does not become acceptable. The creation of new disciplines requires the setting up of new networks of communication, and is a particularly interesting field of sociological and historical study.

In its most radical form, the sociology of science has done far more than uncover the operation of scientific networks. It has sought to extend the realm of the 'sociology of knowledge' to include the natural sciences. The sociology of knowledge was originally developed to explain how cultures attempt to legitimize their philosophical and religious beliefs by claiming that they embody absolute truths about the real world. In its original form, science was excluded from the project on the assumption that it alone really did give access to the truth. Now sociologists of science have begun to argue that there is no reason why science should be so arbitrarily excluded – why its claim to give access to the truth is accepted when all others are rejected. The 'strong programme' in the sociology of knowledge claims that the structure of scientific knowledge can be explained by noting its congruences not with nature, but with the social interests of those who promote it (Bloor, 1976; the strong programme is often associated with the 'Edinburgh school' based on the Science Studies Unit at the University of Edinburgh). Whatever the practical success of science, the fact that its theories offer only models of reality must leave open the possibility that the models have been constructed to reflect the values of the social groups whose interests are best served by the promotion of these particular models. As in other areas, the claim to have absolute knowledge of reality is a device used to gain influence over others. 'Knowledge' represents no more than a collection of opinions that people can be persuaded to accept as the truth, and this is as much the case for science as for any other alleged source of knowledge.

The strong programme takes historians to task for accepting a rigid distinction between the internal and external aspects of science. No one has ever denied that science has effects on the world outside, but it has often been assumed that this is a one-way process. Science discovers the truth, and society then has to adjust to the consequences if (as in the case of evolution) the new information violates traditional beliefs. It was only permissible to look for external influences acting *upon* science when trying to explain how earlier generations had been led astray. If the scientists of a particular era made a discovery that we accept as valid, their actions require no explanation other than that they had used the scientific method correctly. Only if they accepted a theory that we now know to be wrong would the historian be willing to postulate religious or other cultural influences that led them to ignore the obvious. The strong programme claims that it is just as necessary to uncover the ideological factors that shape the creation of the theories we still accept as valid today. All theories, defunct or surviving, have an ideological dimension that must be exposed if we are to understand why these particular ideas about nature were proposed.

Most scientists and philosophers of science have reacted angrily against the strong programme, claiming that it opens the doors to a ludicrous relativism in which one theory is as good as any other. Surely it cannot be accepted that a future genetics will reject Mendel's laws or the double helix model of DNA? Supporters of the strong programme reply that theories only succeed in the scientific arena if they can be exploited successfully in the control of nature. Their success so far does not exclude the possibility that other theories might do just as well, offering parallel but different explanations of the same phenomena. The occurrence of scientific revolutions in the past must surely lead us to expect similar paradigm shifts in the future which will expose the inadequacy of our present 'knowledge'. In fact the original understanding of Mendel's laws had to be significantly modified by later geneticists, and the concept of the gene has undergone major transformations in the course of the twentieth century. Bloor (1976, ch. 6) even explores the possibility that there might be an alternative mathematics, pointing out that some mathematical truths that we now take for granted were rejected by the ancient Greeks (e.g. the concept of one as a number).

Inevitably the strong programme draws many of its examples from the history of science, and urges all historians to use its methods in trying to understand past developments (Barnes and Shapin, 1979; Shapin, 1982). Pioneering efforts in this direction had already been made from a Marxist perspective in Robert Young's analyses of the nineteenth-century evolutionary debates (collected in Young, 1985). Adrian Desmond (1982) has explained complex developments in late-nineteenth-century palaeontology in terms of professional and ideological pressures, while – more directly relevant to our present topic – Donald Mackenzie (1982) has applied the same techniques to the biometrician–Mendelian debate of the early twentieth century. Considerable excitement has been aroused more recently by Martin Rudwick's (1985) attempt to show that major subdivisions of the stratigraphical column were not discovered but were arrived at by negotiation among competing schools of nineteenth-century geologists. Opposition continues, however, and it has been argued that it will be much more difficult to confirm the role of nonempirical factors in the experimental sciences, where theories must be constantly tested against new phenomena.

Most historians have little interest in defending the old view that science offers completely objective knowledge, but at the same time they find it difficult to exploit the sociological perspective with complete satisfaction. One common criticism centres on the tendency for exponents of the strong programme to slip into an almost determinist view of theory-creation. If theories are modelled on the society within which they must function, does this not imply that there will be a single theory which most appropriately embodies those values? (The classic example is natural selection, which is often portrayed as a direct projection onto nature of the ruthless attitudes of Victorian capitalism.) Some early versions of the sociology of science certainly seemed to imply this kind of determinism, and were rejected as presenting a grossly oversimplified picture of science's development. Modern sociologists of science vigorously deny that their position requires a determinist view of how theories emerge. They point out that the establishment of a successful new theory involves the complex and often unpredictable interaction of a host of professional and ideological pressures. Which theory actually comes out on top may often depend on little more than luck, for instance in attracting persuasive or politically sophisticated adherents.

Yet some historians (myself included) suspect that the initial tendency is for the sociologist to pick out a single nonempirical explanation, and only admit a greater complexity when challenged by historians. The problem is that if theory construction is not socially *determined*, then it becomes virtually impossible for the historian to feel certain that he or she has correctly identified a less rigorous influence. It is relatively simple to propose interesting parallels between theories and social values, but much less easy to chart such influences through the complex details of historical (or modern) debates. The value of the sociological explanations is also undermined by the fact that the response of a particular individual or group may be influenced by both professional *and* broader political interests, which do not necessarily coincide and may even conflict with one another. The sociologist can thus move back and forth between the two levels of explanation, according to which best suits the case under review. Critics suspect that their method (like that of the pseudo-sciences) can be twisted to explain practically anything.

These remarks are intended to show that the sociological approach will not be followed uncritically in the rest of this book. But one does not have to accept the whole of the strong programme to believe that the orthodox image of science is woefully one-dimensional. Philosophers, historians and sociologists have all in their different ways challenged the claim that science advances solely by the piling up of factual discoveries. The genie is out of the bottle, and we shall never again return to a simple acceptance of the scientists' own claims about how their disciplines have originated. One does not have to adopt a sociological perspective to recognize that the image of science as a sequence of great discoveries is part of the mythology that sustains the scientific enterprise. The notion of the heroic discoverer who made the breakthrough that initiated each new phase of development is something that originated along with the modern form of science itself (Schaffer, 1986). The orthodox account of the growth of any science is inevitably a form of Whig history – history rewritten by the winners to ensure that the past seems an inevitable prelude to the present. Political historians have long recognized that we need to look behind the façade of accounts written by those who have an interest in adapting the past to their own purposes. Historians of science have now learned the same kind of caution, and this book is an attempt to survey the results of their enterprise in the area of genetics.

2

Heredity before Darwin

At a practical level, people have always had an interest in heredity. Parents were aware of the extent to which their own physical and mental characteristics were (or were not) transmitted to their children. Animal breeders kept track of valuable blood-lines in horses and other domesticated species. Yet the subject is so complex that generalizations were hard to come by. Clearly there was a tendency for characters to be inherited, but there seemed many exceptions and the field was riddled with superstitions and old wives' tales. Naturalists inevitably took an interest in the reproductive process and offered speculations that bore upon the question of heredity. The most obvious question that confronts the historian looking at the pre-Mendelian period is: to what extent did these early ideas anticipate the findings of modern genetics?

This is exactly the kind of question that would be asked by a geneticist who took an interest in the period before the discipline emerged in its modern form. The professional historian of science, however, suspects that this kind of question has been designed to obscure rather than clarify our understanding of the past. If, as outlined in the previous chapter, we accept that earlier conceptual schemes were different to those adopted by modern scientists, then any attempt to view past developments merely as stepping stones towards the present will produce a distorted image of what was really going on. If we wish to know why Mendelism revolutionized modern biology, we must try to understand why earlier theories tended to bypass issues that we now find of crucial importance. This can only be done by trying to reconstruct how those theories functioned within the intellectual climate of the time.

The orthodox approach to the history of genetics treats the pre-Mendelian period as a scene of dark confusion, against which a few bright luminaries appear from time to time. These are the precursors of Mendel – earlier naturalists who seem to have been almost on the point of discovering the true laws of inheritance. Precursor-hunting is characteristic of the scientists' own approach to the history of science. Since the orthodox model of science recognizes a series of great discoveries that laid the foundations of modern knowledge, it seems only reasonable to suppose that a few individuals might almost have anticipated each step in the process. If the true discoverer did no more than clarify a point that is obvious to any impartial observer, then surely some earlier students of the field must also have come close to the truth. It is easy enough to go through the literature of earlier periods trying to pick out isolated passages that seem to anticipate Mendel's findings. Various figures can then be hailed as the precursors or forerunners of Mendelism. The whole history of the subject is reduced to a catalogue of great discoveries, with a series of footnotes giving credit to those who almost got there but didn't quite make it.

Historians of science have been forced to abandon precursor-hunting if they wish to make a serious effort to reconstruct how earlier thinkers viewed the natural world. Instead of picking out isolated passages which, taken out of context, seem to anticipate modern ideas, we try to decipher the context within which the earlier theories and observations were discussed. To do this we must be willing to take quite seriously a collection of ideas that seem ludicrous to the modern scientist. Paradigms are always incompatible with one another, and it is simply impossible to evaluate one by the standards of its successor. If we really want to know how science developed, we must be able to think ourselves into the earlier paradigms so that we can understand the debates by which new paradigms were introduced. This will require us to follow the intricacies of theories that are no longer accepted and complex research programmes that modern science dismisses as totally misguided. But only then will we know what the science of earlier times was really like, and thus be in a position to ask meaningful questions about how modern theories replaced the old ones.

Application of this technique to the field of heredity leads inevitably to the conclusion that earlier biologists were not unconsciously striving to distinguish Mendelian truth from pre-Mendelian superstition. In the course of time a number of alternative theories were explored which bore upon the question of inheritance, and none of these contained anything like the model of character-transmission developed by the Mendelians. Few naturalists saw the study of heredity as a distinct field of investigation. The phenomenon of inheritance was viewed from the context of theories that covered a much wider range of topics. The crucial question for most natural-ists was that of generation or reproduction: how is the new organism constructed? To their eyes it was vitally important to understand how an embryo grew from material supplied by its parents. The question of how and why particular characters are transmitted from one generation to the next was incidental to this much broader issue. Even when the question of heredity was addressed directly, the results were interpreted in terms of theories that would make it impossible for anyone to believe in the existence of hereditary units transmitted unchanged through a series of generations. The crucial issue in the history of genetics is: how did the notion of fixed units eventually emerge? Only by understanding how revolutionary that concept was can we hope to appreciate the significance of Mende-lism's role in shaping modern biology.

When inheritance is subordinated to the question of how the new organism is formed, there is an instinctive tendency for the naturalist to visualize the parents' contribution as a process of *manufacturing* an entity that will become their future offspring. Heredity is a process by which something made by the parent is transmitted to the offspring – *not* the transmission of characters that can be thought of as existing independently of how the new organism is formed. The parents' characters are somehow im-printed upon the egg or semen from which the embryo is formed, so the offspring in a sense 'remembers' the structure of its parents' bodies. Inheritance links only the two adjacent generations; the transmission of a particular character over a series of generations occurs only because each distinct act of reproduction tends to duplicate the parental structure.

In the eighteenth century, some naturalists became so concerned over the problem of generation that they adopted a solution in which individual development did not occur at all. They found it difficult to believe that a living body could actually manufacture the materials from which a new body (its offspring) could be formed, so they supposed that the embryo grows from a preformed miniature created originally by God. This 'preformation theory' has been ridiculed by modern biologists because it seems obvious to us that the embryo is formed by a process far more complex than the expansion of pre-existing parts. The fact that such a theory was once taken seriously by eminent biologists suggests that their conceptual scheme was completely different to ours. For them, generation could only occur through the parent organisms building a copy of themselves. If this was impossible (because no physical theory seemed able to explain how so complex a process could occur) then divine creation was the only remaining explanation of how organic structures are formed. In a sense, the preformationists were quite correct in their realization that no theory could cope with the job of explaining how living structure could be built up afresh in each new organism. They sensed that *something* independent of the parents' own bodies must carry a programme for the construction of each new generation. Given the limitations of chemical and physiological theories at the time, however, they could only imagine that something to be a miniature created originally by God. Their dilemma was, in fact, an early manifestation of the problem that would eventually destroy the theory of reproduction by individual acts of generation.

The preformationists and their opponents shared a common belief that if generation occurred it must represent, in effect, an act of creation by the parents' bodies. This would involve far more than the duplication of existing genetic material: each organism would somehow have to manufacture particles that would copy or remember the structure of the body that had formed them. For those who believed that nature was capable of such complex activities, the most obvious approach was to assume that each part of the body creates particles that remember its own structure, all the particles then being conveyed to the sexual organs for the purpose of reproduction. The material in the egg or semen would thus be a complex mixture of particles derived from the whole body. Even the preformationists tended to assume that individual peculiarities are inherited because the parents' bodies contribute nutritive particles that are absorbed into the growing embryo. This atomistic view of the reproductive tissue naturally encouraged the belief that the parental contributions are mixed to form the structure from which the embryo grows. Each organ in the embryo will thus contain some particles derived from each of its parents. In most cases the offspring will tend to be a blend of the parental characters – if the parents differ in any respect, the offspring will be intermediate between them. This is obviously not the case with sex (or we would all be hermaphrodites), but it seems to occur often enough in all other characters to support the belief that inheritance works by blending and is not particulate in the Mendelian sense.

Equally remote from modern genetics was the belief that changes affecting the parents can be reflected in the offspring. If the parent's body *manufactures* the particles from which the body of its offspring will be constructed, each particle copying the structure of the organ that produced it, then it seems obvious that changes affecting the parent's body ought to be transmitted to the offspring. New characteristics acquired by the parent in response to new habits, a changed environment, or even accidental mutilation, can be expected to appear in the offspring. This 'inheritance of acquired characters' was accepted by the majority of naturalists through into the nineteenth century, and was incorporated into J. B. Lamarck's pre-Darwinian theory of evolution. Again, it was easy enough to find evidence from hearsay and old wives' tales to support the theoretical prediction. The effect that would eventually come to be known as 'Lamarckism' only became a biological heresy after the modern concept of the gene introduced the view that the parents are merely vehicles for transmitting fixed characters from one generation to the next.

Given the prevailing belief in the fixity of species, the majority of naturalists wanted a theory of generation that would restrict variation within narrow limits. The central purpose of heredity must be to ensure that the next generation will be constructed in the likeness of its parents. Variation within the species arose because external factors tended to interfere with this copying process. A few radical thinkers realized that if parental variations can be transmitted, then in principle there is no limit to the degree of change that may build up within a species over a long period of time. But most pre-Darwinian naturalists wanted to avoid this implication, and introduced factors into their theories designed to ensure that inheritance will produce an accurate copy of the basic parental form. Even Darwin still saw inheritance and variation as essentially antagonistic processes. Only in the late nineteenth century did biologists begin to suspect that heredity and variation were merely different aspects of the same process, due to the constant circulation within a population of genetic units corresponding to different forms of the same character.

Materialism and Generation

Throughout the period before 1700, the prevailing view of life tended to be based on what later writers would call 'vitalism'. In other words, it was commonly assumed that the living body was more than a material system: it had to be driven or vivified by a non-physical force that somehow disappeared at the death of the organism. Modern biologists tend to sneer at vitalism, dismissing it as little more than a form of superstition designed to impede scientific (i.e. materialistic) study of life. Historians of science now take a much more sympathetic approach to the relationship between vitalism and materialism. There were many problems that at first seemed insoluble in materialistic terms, given the limited sophistication of the physical sciences at the time. It may be no bad thing for biologists to hypothesize a non-physical force to explain processes beyond their comprehension, if in so doing they can introduce some order into their observations of living things. Although the long-term trend of biology has been towards the creation of materialistic explanations, this trend has constantly been interrupted by episodes in which the preliminary investigation of particularly refractory problems had to be conducted within a non-materialistic conceptual framework.

The limitations of materialism seemed most evident in the area of generation or reproduction. It was difficult enough to believe that purely physical forces could sustain the life of an adult organism – how much more unreasonable it must have seemed to propose that those same forces could actually build a new living body out of unorganized matter. In the ancient world, Aristotle had supposed that a non-physical influence derived from the male semen was the active power that shaped the growing embryo from the female menstrual blood. (For information on early theories see Cole, 1930; Needham, 1959; Stubbe, 1972.) Such a view had obvious implications for heredity, since in principle it gave the father the only active role in the creation of the offspring. However, Aristotle held that, in practice, the menstrual blood can 'interfere' with the power of the male's influence to allow the transmission of some maternal characters. Other ancient writers held that both male and female produce a semen which, when mixed in the womb, forms the basis from which the embryo grows. This allowed a more balanced view of male and female influences, although only the more radically materialist philosophers of the ancient world claimed that the embryo was formed solely from the material substance of the semen.

Ancient materialism had been driven underground during the medieval period, but similar ideas resurfaced during the 'scientific revolution' of the seventeenth century (Dijksterhuis, 1961). The new science of mechanics pioneered by Galileo, Descartes and Newton created a sense of optimism that led more radical thinkers once again to argue that the laws of material nature are sufficient to explain the behaviour of the whole universe. This 'mechanical philosophy' inevitably set itself the goal of accounting for the functioning of the living body in materialistic terms. According to its most eminent exponent, the French philosopher and scientist René Descartes, the animal body was nothing more than a complex machine. All signs of 'mental' activity are merely indications of processes going on in the nervous system. The dog yelps when it is kicked, not because it feels pain, but because yelping is a warning of damage to the system – just like a modern computer flashing a red light to indicate a malfunction. Only in human beings is the activity of a material body somehow linked to a soul capable of mental activity. With this exception, Descartes' 'animal machine' doctrine proclaimed that scientific biology would proceed by reducing itself to a branch of physics.

Descartes' mechanistic philosophy became a basis for the radical materialism of the eighteenth century (Vartanian, 1953). But its success was always limited both as a philosophy and as a scientific research programme. Conservative thinkers preferred to believe that the structure of the material universe had been established by its Creator, even if it was sustained by nothing more than the laws of physics He had endowed. There were, in any case, obvious limitations to the application of Descartes' principles in biology. With no sophisticated theory of chemistry, a serious assault on the problems of physiology was impossible, and vitalism could not be excluded from the field. The question of generation was even more critical: to many naturalists it seemed ludicrous to suppose that a living body, conceived as nothing more than a complex piece of clockwork, would be able to manufacture another in its own likeness. William Harvey – whose work on the circulation of the blood had been used by Descartes to support the idea of a machine-like body – endorsed Aristotle's view that a non-physical influence from the father somehow triggered off the growth of the embryo, although he held that the mother's contribution is encapsulated in an ovum or egg. Harvey's *De Generatione Animalium* of 1651 (translated in Harvey, 1965) also supported Aristotle's claim that the embryo grows by 'epigenesis' or the sequential production of its various parts in the course of time. It seemed impossible to believe that any material structure could store the information needed to control such a complex sequence of events.

By the end of the century, however, the newly discovered microscope was leading some observers to question the validity of epigenesis. Microscopists such as Malpighi showed that parts of the embryo often began to form long before they became visible to the naked eye – and it was possible to hypothesize that they were there before even the microscope could detect them. This possibility offered a way out of the dilemma facing those naturalists who supported the mechanical philosophy but could not see how it might apply to generation. They now began to argue that the embryo was actually preformed within the ovum. In other words, a complete miniature organism was already in existence at the very beginning of the growth process, so the growth of the embryo consisted of no more than an expansion of preformed parts. From a mechanistic perspective, this had the advantage of simplifying growth to manageable proportions: it was easy enough to suppose that physical forces could incorporate new material into a previously existing structure so that it would grow in size. From this point through until the last decades of the eighteenth century, the majority of naturalists accepted that the embryo grows not by epigenesis but by simple expansion (for details see Adelmann, 1966; Bowler, 1971; Gasking, 1967 and Roger, 1963).

This 'preformation theory' is now frequently ridiculed for its blatant refusal to acknowledge the obvious fact that the embryo undergoes a complex process of development that is no mere expansion. But even more scorn is heaped upon the most extreme version of preformationism, more properly known as the theory of pre-existing germs. According to this theory, it was not enough to suppose that a preformed miniature existed only within the fertilized egg, since this would still leave the problem of explaining how fertilization – if it were only a physical process – could build the miniature out of particles supplied by the parents' bodies. No, the miniature must have existed in the egg *before* fertilization. Indeed, since no physical process could build a living body out of disorganized matter, the miniature must have been in existence before even the mother was born. In its final version, the theory of pre-existing germs held that all organisms grow from miniatures or 'germs' created by God at the beginning of the universe, stored up one within the other like a series of Russian dolls. The first woman, Eve, literally contained within her ovaries the whole of the rest of the human race, generation after generation of miniatures packed one inside the other, each waiting for an act of fertilization to give it a chance to grow. On this model, the male semen merely provided the stimulus that triggered off the expansion of the outermost miniature. A rival version of the theory held that the germs were stored within the spermatozoa of the male semen, but this was never very popular, given the vast wastage of sperms in every sexual act. Most eighteenth-century naturalists, in fact, dismissed the spermatoza as mere by-products of the seminal fluid.

The theory of pre-existing germs sounds so bizarre to modern ears that we can hardly blame biologists for supposing that it was only promoted for non-scientific reasons. Yet it remained popular through most of the eighteenth century, and was accepted by many biologists who are still respected for their work in other fields. It thus behoves us to look more carefully at the theory and its implications for the study of heredity. Certainly, there were non-scientific factors at work. Those who supported the theory belonged to what might be called the conservative wing of the mechanical philosophy. They accepted that the material universe was governed solely by natural law, but wished to retain the belief that God was directly responsible for the creation of each human being (and by implication each animal and plant). The Swiss naturalist and philosopher Charles Bonnet even twisted the concept of pre-existing germs into an explanation of the resurrection of the body promised by traditional Christianity (Anderson, 1982; Bowler, 1973). But this was no mere Christian fundamentalism. Bonnet, like many eighteenth-century thinkers, was desperately anxious to show that the human mind can comprehend the order of nature. By picturing God as the creator of both human souls and the germs of all living creatures, Bonnet ensured that the world will be accessible to our senses – not just a bewildering muddle.

Bonnet's support for pre-existence thus came from motives that a modern scientist ought to sympathize with: the belief that material nature is, despite its complexity, open to human investigation. His repudiation of the (to us) obvious fact that some structures only appear later in the growth of the embryo must be judged in terms of the excitement generated by the microscope, which made it difficult for anyone to claim that the rudiments of a structure were not already present before it became visible. (Only later in the century was it possible for more careful observations to show that these rudiments did not exist, and that new structures actually developed in the growing embryo.) In fact Bonnet did not believe that a complete miniature organism lay within the egg, only a 'network' that would become visible as the parts were filled in by nutritive matter. There was no potentially visible human being in the ovum – no more than a man-shaped balloon would be recognizable when it is deflated and folded up. It is thus unwise to accuse Bonnet of simply ignoring the results of observation – indeed he had gained his early reputation through his discovery of parthenogenesis in aphids, and only later turned to the philosophical implications of generation.

Other supporters of the theory of pre-existing germs are also known for their contributions to 'orthodox' science. Albrecht von Haller is still noted for his contributions to physiology, although he led the defence of pre-existence in the later decades of the eighteenth century. Lazarro Spallanzani is remembered for his demonstration that spontaneous generation does not occur – an important step forward in modern terms, yet a vital line of support for pre-existing germs as far as Spallanzani himself was concerned. Spontaneous generation (the creation of living organisms directly from inert matter by natural processes) was anathema to Bonnet and Spallanzani precisely because it seemed to lead the materialist into a morass of speculation in which nature became a chaos of indecipherable phenomena (Mazzolini and Roe, 1986). They preferred to believe that life came only from germs created by God, thereby guaranteeing the fixity of species and the stability of the natural world. What looks like a 'modern' discovery thus had a very different context in the eighteenth century (we shall return to the question of spontaneous generation below).

But what did the concept of pre-existing germs imply in the field of heredity? In its most extreme form, the claim that all organisms grow from germs originally created by God makes nonsense out of any attempt to trace individual characters from one generation to the next. If the germ defines the individual in all his or her peculiarities, then the organism's structure depends on what God created, not on any transmission from the parents. In fact a considerable debate raged around the question of whether God deliberately created monstrosities and deformities (some of which reappear in several generations). Most naturalists agreed that God was *not* directly responsible for such occurrences – they were the accidental results of defects in the absorption of material into the expanding germ. In his *Considérations sur les corps organizés* of 1762, Bonnet declared that the germ defines only the species to which the organism will belong, not its individual characteristics (Bonnet, 1779, 6: 392–3). This solved the problem of maintaining natural order, but left the details of growth which define individual peculiarities to natural processes. Inheritance of parental characters was possible because the germ grew by absorbing nourishment first from the father's semen and then from the mother's womb. Since the parents' bodies controlled the flow of material to their reproductive organs, individual characteristics could be passed on to the offspring. Bonnet noted a number of cases where a particularly obvious character had been transmitted through several generations of the same family.

If the preformation theory violates our modern view that the embryo is not simply folded up within the female ovum, it must be remembered that the alternative was to suppose that all the responsibility for constructing the origins of the embryo rested on the generative power of the parents' bodies. No one at this time could think in terms of chemical molecules that could both replicate themselves and store the information needed to control a complex sequence of biological events. In a sense, Bonnet's idea of preformed rudiments is about as close as the eighteenth-century mind could get to the modern view that material entities (the genes) hold the key to the development of characters in the growing embryo. It is significant that the issue of preformation versus epigenesis resurfaced in a different context during the last decades of the nineteenth century, when the origins of the modern concept of the gene were first being sketched in. August Weismann's hypothetical substance of heredity was called the 'germ plasm', harking back to the notion of a material entity, the germ, from which the embryo had once been supposed to grow. The claim that adult characters were defined by material structures in the cell nucleus was known as 'nuclear preformation' because it seemed to recall the old belief that the essence of the embryo was already determined before growth itself began. Given that the only alternative to preformation in the eighteenth century was the view that the new organism was created from material generated by the parental bodies, Bonnet's theory was neither more nor less 'modern' than that of his opponents, since both sides accepted a conceptual system that would be unthinkable today.

The alternative to the theory of pre-existing germs may at first sight appear to anticipate modern ideas of heredity and development, but the resemblance is only superficial. Although 'epigenesis' is normally presented as the opposite of 'preformation', very few opponents of the preformation theory actually believed in a sequential development of parts in the embryo. They too saw the embryo expanding from a complete miniature, but believed that this structure was formed during the act of fertilization from material supplied by both parents. When true epigenesis was suggested, no explanation of the successive production of parts could be offered. The physical sciences of the eighteenth century simply could not supply a theoretical model sophisticated enough to explain how a very complex sequence of events could be 'programmed' into a material structure. Whenever organic processes of this complexity were hinted at, they were taken as an excuse for qualifying the mechanical philosophy by hinting at hylozoism (the belief that matter itself is alive). A material explanation of organic activity could only be advanced by attributing the purposeful behaviour of life itself to the ultimate particles of matter. Widespread support for true epigenesis only came about towards the end of the century, when the principles of materialism were self-consciously repudiated.

In the meantime, an ambitious, if sometimes rather woolly, materialism became a powerful tool in the hands of more radical thinkers determined to challenge orthodox ideas and established conventions. The eighteenth century was the 'Age of Enlightenment' when many thinkers proclaimed the power of human reason to sweep away ancient superstitions (Cassirer, 1951; Gay, 1966–9; Hampson, 1968). Materialism could be used to assault traditional religion, and hence to undermine the social order that the churches sustained. Deists reduced the Creator to an abstract figure with little concern for human affairs, while atheists such as the Baron d'Holbach maintained that there was no divine plan of nature at all. Theories of generation were of crucial importance in the debate between the conservative and radical wings of eighteenth-century thought. Pre-existing germs implied both the existence of God and the stability of nature. By contrast, any theory in which material nature was itself responsible for the generation of living things seemed a recipe for atheism and chaos. Deists and atheists were anxious to develop theories that would challenge preformation by explaining how the new organism could be formed from particles supplied by the parents in the sexual act. To drive home the message even more effectively, this approach was often extended to include the hypothesis of spontaneous generation. In some circumstances, nature might be able to dispense with living parents altogether and build new organisms directly from inorganic matter. Nature became a haphazard world of monsters and new species, thereby undermining the whole basis of the 'divine plan'.

From the perspective of natural philosophy, the new approach to generation could be presented as an attempt to formulate a genuine explanation of how living things were formed. One of the most common complaints against the theory of pre-existing germs was that it evaded the issue by pretending that God was the only source of organic complexity. This point was emphasized by Pierre Louis Moreau de Maupertuis in his popular *Venus physique* of 1745 (translation 1968). Maupertuis used Harvey's observations to support the claim that the embryo grows by epigenesis: some parts appear later than others, so the whole structure cannot have been preformed in the ovum. He revived the twin-semen theory, arguing that both parents contribute a fluid semen which, when mixed in the mother's womb, forms the basis of the embryo's structure. Each part of the adult body forms particles that are somehow responsible for establishing the corresponding part in the offspring. The particles are concentrated in the seminal fluid, and since the embryo is formed only after mixture, both parents help to determine the character of their offspring.

Maupertuis' attack on pre-existence was confident enough, yet his theory offered no convincing explanation of how the seminal particles are guided to the appropriate part of the growing embryo. There was certainly no mechanism to explain why some parts of the body were formed later than others, as implied by Harvey's work. Maupertuis hinted that a force resembling gravity might draw the seminal particles together to form the embryo. As to how the particles 'knew' their own place in the overall structure, the *Venus physique* was silent. Maupertuis' *Système de la nature* of 1751 postulated a faculty resembling memory by which the particles knew which part of the body they had been derived from. He also implied that they had an active desire to seek the appropriate place in the growing embryo. Here was the great paradox of the materialist approach to generation: without an adequate conception of molecular complexity, it was only possible to explain so complex a process by attributing the properties of life to matter itself.

Maupertuis is included in most of the orthodox histories of genetics because he also described observations in which the inheritance of particularly obvious characters was traced through a number of generations. His account of the inheritance of polydactyly (the possession of a sixth digit on hand or foot) has led some historians to describe him as a precursor of Mendelism (e.g. Glass, 1959a). Yet it is clear that Maupertuis' explanation of his results bears no resemblance to modern genetics (Sandler, 1983). He had no conception of a hereditary unit coding for polydactyly, transmitted through a series of generations. Instead, he supposed that the character originated in some malfunction within the formative process of a particular embryo. Once present in an adult body, the character might be passed on because the semen of that individual would contain particles that 'remembered' the new configuration. If these particles happened to predominate when male and female semen was mixed, the offspring would inherit the character. But if the offspring did *not* acquire the character, it could not transmit it to future generations, since Maupertuis' theory required the parents' bodies to manufacture seminal particles 'in their own image'. Characters were not transmitted independently from one generation to the next.

A similar theory of generation underlies the thinking of one of the eighteenth century's greatest naturalists, the Comte de Buffon. Buffon's massive *Histoire naturelle* began publication in 1749 and eventually included descriptions of all the higher animals, along with theoretical speculations that are popularly supposed to have anticipated the modern concept of evolution (Wilkie, 1956; Lovejoy, 1959; Farber, 1972; Bowler, 1973). Buffon was searching for a Newtonian explanation of how the world came into its present state, and he shared Maupertuis' view that preformation simply evaded the crucial issue of the origin of organic structure. His own theory of generation (in Volume 2 of the *Histoire naturelle*) followed Maupertuis in postulating 'organic particles' concentrated in both parents' semen, and from which the embryo of the new organism is formed. Significantly, Buffon seems to imply that the outline of the complete embryo is formed soon after the mixture of the semen: he did not really believe in epigenesis.

Unlike Maupertuis, however, Buffon tried to avoid giving the organic particles the properties of life itself. Instead he introduced the concept of an 'internal mould' which somehow directs the particles to the correct place in the matrix of the new organism. The internal mould soon became a guarantee of the fixity and permanence of species (Buffon's transformism allowed only for limited changes within the established types of animal life). In effect, the mould served the same purpose as Bonnet's series of pre-existing germs, and its true nature was never fully defined. One suspects that Buffon sensed his inability to provide a truly materialistic explanation of so intelligent and pervasive an entity. We are simply left to assume that there are fundamental constraints operating within nature, forcing the organic particles to conform to certain types of structure defining the animal species. At the level of individual reproduction, Buffon could allow for the inheritance of particular characters by supposing that the peculiarities of the parent's structure would influence the organic particles in its semen and thus have a chance of being transmitted to the offspring. He also believed that any change in the environment would affect the reproductive systems of all the individuals exposed to it, producing definite changes in the next generation. Over a long period of time, this process would lead to significant differences between geographical races of the same original type.

As part of his campaign against the theory of pre-existing germs, Buffon supported the experiments on spontaneous generation performed by John Turberville Needham (1748; see Roe, 1983, 1985; Mazzolini and Roe, 1986). These experiments seemed to reveal that sterilized meat gravy kept in a sealed vessel could nevertheless generate micro-organisms, presumably by the free organic particles coming together under natural forces. Spallanzani would eventually show that Needham had not sterilized his materials adequately, so his micro-organisms had actually grown from spores, but in the meantime the experiments were widely taken as evidence that nature had the power to generate life from nonliving matter. This claim was exploited by atheists like d'Holbach as the basis for their attack on the traditional belief in divine creation. If micro-organisms could be generated in the laboratory, more complex animals might be produced directly by spontaneous generation under certain (as yet unknown) natural conditions. Buffon himself used this idea in later volumes of the *Histoire naturelle* to explain the origin of life on the earth, although he specified rigid limits upon nature's allegedly creative powers. His concept of the internal mould allowed him to suppose that the new forms of life would always conform to one or another of the basic types that define the known animal forms.

Needham himself strove to dissociate spontaneous generation from the link with atheism, but his critics were only too well aware of how the concept was being used by the more radical Enlightenment philosophers. D'Holbach's *Système de la nature* (published under an assumed name in 1770) was known as the 'Bible of atheism' since it treated religion as a fraud designed to maintain an outdated social system. D'Holbach had studied chemistry, and for him spontaneous generation was no more than a glorified chemical reaction by which living organisms were formed from inert matter, governed only by the natural affinities of the material particles. Unlike Buffon, however, d'Holbach would impose no limits on the power of nature's creativity. He was determined to show that the structure of the material world exhibits no traces of a divine plan. Nature *experiments* with the production of new forms of life: even in normal reproduction, the frequent appearance of monstrosities and deformities confirms that the generative powers are not restricted to known types. The same is true for spontaneous generation, so that nature is constantly producing new forms of life, of which only the more viable types will actually survive. D'Holbach supposed that monstrosities might sometimes be able to reproduce themselves and thus found new species, while the ever-changing environment will constantly modify existing species by interfering with their reproductive processes. Here the new theory of generation became the basis for a vision of nature as a creative entity in a constant state of flux, with trial-and-error experimentation replacing the stability of divinely created species. The last thing that d'Holbach would have wanted was a theory in which inheritance is governed by units

that can maintain their characters unchanged over many generations.

As we shall see in the next section, the materialist approach to
generation was largely abandoned in the last decades of the century.
But one naturalist of this later period helped to ensure that certain
aspects of the old conceptual scheme would not be forgotten. J. B.
P. A. de Monet, Chevalier de Lamarck, is widely known as the
author of one of the first comprehensive theories of biological
transformism or evolution (Gillispie, 1959; Hodge, 1971;
Burkhardt, 1977; Mayr, 1972; Jordanova, 1984). This is not the
place for a discussion of Lamarck's overall world view; suffice it to
say that the one concept for which he is remembered formed but a
small part of his thoroughly pre-Darwinian (and pre-Mendelian)
approach to biology. Nowadays 'Lamarckism' means the inheritance of acquired characteristics, although that concept was not
introduced by Lamarck and was in fact a typical product of the
eighteenth-century approach to generation. By incorporating the
effect into his highly controversial theory of evolution, Lamarck
ensured that a thoroughly pre-Mendelian view of heredity would
remain a talking point throughout the rest of the nineteenth
century.

The best-known account of Lamarck's theory is contained in his
Philosophie zoologique of 1809 (translation 1914). Unlike Buffon,
whose son he had tutored, Lamarck provided no comprehensive
theory of reproduction, although he retained the idea that spontaneous generation provides the ultimate source of all living things.
Lamarck insisted that species are not fixed; they are transformed in
the course of time as they adapt to changes in the external
environment. The process by which this adaptive transformation
occurs is the inheritance of acquired characters. In a new environment, animals will develop new habits and will exercise their bodies
in new ways. If a part is used more than before it will increase in size
– just as the muscles of the weightlifter's arms are developed.
Lamarck assumed that such 'acquired characters' are transmitted,
at least partially, to the offspring – as though the weightlifter's
children would be born already with slightly bigger than normal arm
muscles. Since the next generation will continue with the new
habits, the corresponding characters will be enhanced gradually
over the centuries until eventually the whole species is transformed.
The classic illustration is that of the giraffe, whose long neck has

been formed as the result of ancestral generations striving to reach the leaves of trees (1914, p. 122).

The inheritance of acquired characters is, of course, repudiated by modern genetics, because we see no mechanism by which the genes can absorb information from the body in which they are enclosed. Yet it has been accepted as a genuine effect since time immemorial (Zirkle, 1946) and even today many non-scientists almost instinctively believe that the effect ought to work. The inheritance of acquired characters originated within a pre-Mendelian notion of heredity that is more in tune with common-sense or folk belief. It was taken for granted by the eighteenth-century naturalists precisely because their theories of generation assumed that the parents' bodies somehow *produce* the material from which their offspring will grow. The parents do not merely transmit characters over which they have no control. Once it is assumed that inheritance is a direct transmission of the parents' adult characters to their offspring, it seems obvious that any changes to the parents' bodies must affect the generative process and may thus be passed on. Such a natural assumption can only be undermined by dismantling the whole conceptual scheme within which it functions. Modern genetics emerged by postulating material entities that could transmit characters independently of the parents' bodies. The eighteenth-century naturalists had certainly thought in terms of material processes linking parent to child, but their materialism was constrained by a view of generation that would not allow them to visualize those processes in modern terms.

The Rise of Developmentalism

Since both sides of the debate over preformation tended to ignore epigenesis, the mechanistic approach to generation had effectively excluded any use of embryology as a model for exploring the theme of development through time. Buffon and Bonnet both accepted that the embryo grew only by expansion; their debate was centred on whether or not an initial act of generation could form a miniature organism from particles supplied by the seminal fluids. In the later decades of the eighteenth century, however, the mechanistic viewpoint came increasingly under fire. Embryology took on a new lease of life as a whole generation of biologists threw themselves into a detailed study of how the new organism is formed. Superficially, it looks as though more careful observation of growth disproved preformation by showing that the embryo does indeed undergo a complex process of development. But the crucial nature of the observations was only admitted by those naturalists who had already repudiated the mechanical philosophy. Since they no longer wanted to visualize the creation of the new organism as a mere locking together of material particles, they could accept the evidence that showed the embryo being constructed by a complex and apparently purposeful process. Observation and theory thus combined to give a new climate of opinion in which the development of the embryo could be seen as a model for the general tendency of nature to advance toward higher levels of complexity.

Strictly speaking, this new climate of opinion tended to deflect attention away from the problem of heredity. The new embryology was conceived as a branch of morphology: its primary task was to understand the structure or form of the organism and to describe how that structure was created from the fertilized egg. The question of why the egg developed in this way was evaded by postulating teleological forces that were not reducible to mechanical processes. Since development was not the unfolding of a programme stored in a mechanical system, there was little interest in how parental characteristics were transmitted to the offspring. Embryologists were, however, drawn to the study of monstrosities, where 'errors' of development might throw light on how the process actually worked. This whole episode is of profound interest to historians of embryology, but is generally dismissed as a backwater by those concerned with the antecedents of modern genetics. Apart from destroying the obvious nonsense of preformationism, the developmentalist philosophy did nothing to further the study of heredity; indeed it actively discouraged such studies by falling back on the vitalist concept of a purposeful growth force.

The embryologists were not, as we shall see, simple-minded vitalists, although their fascination with development certainly allowed teleology to regain its place in biology. Even so, their destruction of preformationism helped to define the problems that later materialists would have to confront in the area of reproduction. If parental characters were transmitted to the offspring by a material process associated with the sexual act, the material agent of transmission would have to be complex enough to include a

programme capable of controlling a whole sequence of developmental stages. Developmentalism also profoundly influenced two
areas of biology that would help to shape the late nineteenth
century's renewal of interest in heredity. The theory of evolution
provides a classic example of how the growth of the embryo was
used as a model for progressive developments on a wider scale.
Darwinism turned attention once again to the question of inheritance, but was at first enmeshed in pre-Mendelian concepts that
prevented its hereditarian implications from being fully realized.
Embryology also contributed to the growing awareness that the
cellular structure of living tissue offered the key to a better
understanding of its functions, thereby providing the framework
within which the post-Darwinian biologists would try to resolve the
now central problem of variation and heredity.

The new assault on preformationism was begun by the German
embryologist Caspar Friedrich Wolff, whose *Theoria generationis*
of 1759 included detailed observations showing that the parts of the
embryo are constructed one after another from developing tissue.
Wolff was challenged by the noted Swiss biologist Albrecht von
Haller, who stepped in to defend the theory of pre-existing germs
(Roe, 1981). The fact that Haller refused to acknowledge the
significance of Wolff's observations indicates that the debate
centred on fundamentally different philosophies of nature. In such
circumstances there can be no crucial observation or experiment
that will settle the matter once and for all. However convinced we
may be that preformationism was wrong, we must accept that the
theory was destroyed by changing attitudes, not by the piling up of
facts that it could not explain. Far from being hailed as a hero of
science, Wolff had to move to Russia to gain academic preferment.
Only towards the end of the century did his views become more
generally accepted, although he died in 1794.

Wolff could support epigenesis because he no longer thought of
the embryo as a structure that was built by merely slotting together
particles derived from the parents' semen. He believed that each
step in the process depended on the one before: once it was formed,
each part of the embryo secreted new material from which the part
next in the sequence would solidify. This was quite a sophisticated
image of development, but it left open the question of how so
complex a process was controlled. Wolff postulated a *vis essentialis*
or 'essential force' that operated throughout the growth period.

Haller responded by accusing Wolff, in effect, of vitalism – of once again evading the problem of generation by invoking a mysteriously constructive force that existed only for living things. Many later historians have agreed with Haller's assessment, although Wolff insisted that his force was not mysterious and could simply be added to the list of known natural forces. An explanation of generation would only be achieved by recognizing the true character of the process so that forces of the required complexity could be postulated. If these forces did not obey the rules laid down by the mechanical philosophy, this merely showed that the universe was more complex than a piece of clockwork.

Wolff was obviously not a simple vitalist, yet his system still left unsolved the central problem of how the fertilized egg acquired the structure necessary to initiate such a complex sequence of events. On the strength of his later writings, Shirley Roe (1981, ch. 5) suggests that he saw each species as having been endowed by the Creator with a basic 'vegetative' substance capable of growing new individuals with the aid of the *vis essentialis*. This substance seems to have played the same role in his thinking as Bonnet's series of germs or Buffon's internal mould: it somehow defined the basic character of the species. He offered no suggestions about how it was conveyed from one generation to the next and was unable to specify how a material substance could embody the potential structure of a living organism. Nevertheless, he had recognized the central implication of epigenesis, that generation would have to be seen as the unfolding of a potential somehow encoded within the fertilized egg.

As the reaction against Enlightenment materialism developed in late eighteenth-century Europe, Wolff's critique of preformationism at last came into its own. A new generation of German biologists now set out to explore the diversity of organic structures, seeking to understand how the various forms of animals and plants were related to one another, and how the different structures were evolved in the growth process of each type. Historians have tended to associate this renewal of interest in embryology with the *Naturphilosophie* of F. W. J. von Schelling and Lorenz Oken, a mystical philosophy of nature paralleling the romantic reaction against Enlightenment values in the arts. The *Naturphilosophen* believed in the unity of nature and hoped to demonstrate this unity in biology by discovering the underlying ground-plan of which all living species are but variants. They also saw the growth of the individual organism as a process that illustrated the purposeful activity of nature in general. The material world was now seen as merely a projection of a deeper spiritual reality.

Some embryologists at the turn of the century were attracted to *Naturphilosophie* and allowed their work to be used as the basis for an almost mystical developmentalism. But thanks to the work of Timothy Lenoir (1982) we now know that the mystical tradition in German biology was soon overtaken by a much more serious approach deriving its inspiration from the philosophy of Immanuel Kant. The study of individual growth was still central to this programme of 'teleomechanism', but (as the name implies) an effort was made to strike a balance between the search for materialist explanations and a recognition that nature does exhibit purposeful behaviour. The teleomechanists saw growth as an example of purposeful development, but repudiated the *Naturphilosophens'* attempt to portray such activities as manifestations of spiritual forces underlying material nature.

An early move in this direction was made by the eminent naturalist and anthropologist Johann Friedrich Blumenbach. By the 1780s Blumenbach had abandoned preformationism and had postulated a *Bildungstrieb* or constructive force responsible for the development of the embryo. He insisted that this was not a spiritual entity somehow imposing its will on brute matter – the *Bildungstrieb* was certainly goal-directed, but was a force in the normal sense of the term and could exist only in conjunction with material structures. Blumenbach's concept was identified by the philosopher Immanuel Kant as exactly the kind of 'regulative principle' required in the scientific study of biology. Unlike many British scientists at this time, Kant and his followers did not link their teleological approach to an explicitly religious programme. The fact that nature worked in a purposeful way was accepted as a necessary limitation on the scope of mechanical explanations, not as an indication that divine providence was in control of every detail of nature's operations.

Of the anatomists and embryologists who took up the teleo-mechanist programme in the early nineteenth century, two stand out as having promoted images of development that were to interact and conflict with one another through into the post-Darwinian era. In 1821, J. F. Meckel introduced the 'law of parallelism' to describe the growth of the human embryo (Russell, 1916; Temkin, 1950;

Oppenheimer, 1967). He argued that in the course of its development the embryo in effect ascends the hierarchy of animal forms,
passing from the lowest to the highest levels of organization seen
elsewhere in the animal kingdom. The human embryo passes
through phases in which it is first a fish, then a reptile and finally a
mammal. The adults of the lower animals can thus be regarded as
immature versions of humankind, creatures whose growth has been
frozen at a lower level in the hierarchy of potential development.
The study of development had revealed the unifying principle of the
animal kingdom. This announcement was followed by a growing
recognition among palaeontologists that the sequence: fish, reptiles, mammals, represents the order in which the classes were
introduced in the course of the earth's history. The growth of the
human embryo thus came to be treated as a model for the
progressive development of life on earth. This analogy was used
extensively in the early nineteenth century, and was spread to the
English-speaking world by the Swiss-American naturalist Louis
Agassiz (Lurie, 1960). In the age of evolutionism, it would become
the basis for the 'recapitulation theory': the claim that the development of the embryo recapitulates the evolutionary history of its
species (Gould, 1977).

Meckel's image of humankind standing at the head of a linear
scale of development would remain powerful despite the fact that it
had been challenged almost from the start by the work of an
embryologist with a far more enduring reputation, Karl Ernst von
Baer. Also a follower of the teleomechanist programme, von Baer
gained international recognition for his discovery of the mammalian
ovum in 1827. (Some eighteenth-century naturalists had assumed
that the Graafian follicle is the ovum; von Baer showed that the true
ovum is a much smaller body at first contained within the follicle.)
His *Über Entwickelungsgeschichte der Thiere* of 1828 and 1837
contained a major survey of vertebrate embryology. The fifth
scholion of this work (translated in Henfrey and Huxley, 1853)
rejected the law of parallelism on the grounds that the human
embryo is never equivalent to the adult form of a lower animal.
Growth is a process of specialization: the most basic and most
generalized structures appear first, the more specific ones later on.
A human and a reptile embryo may be indistinguishable at an early
stage of development, but each acquires its own specialized
characters in the course of maturity. Von Baer's law itself became

the model for progressionist images of the history of life on earth, but its impact was obscured by the preference of many later naturalists for a linear scheme of development more in line with the law of parallelism (see Chapter 3).

The embryologists of the early nineteenth century had little interest in heredity as such, but their ideas helped to shape the conceptual framework within which Darwin's theory of evolution would be interpreted. Darwin's mechanism of natural selection certainly focused attention on the questions of variation and inheritance, but the developmentalist viewpoint ensured that the questions would be tackled within a pre-Mendelian paradigm. The recapitulation theory soon became linked to Lamarck's inheritance of acquired characters, since both worked best with a model in which heredity was seen as a process analogous to memory. The growing embryo in effect remembers the long sequence of events experienced by its distant ancestors. A biologist who visualized heredity acting in this way would inevitably find it difficult if not impossible to accept the suggestion that characters are transmitted as fixed units from one generation to the next.

Seen from the perspective of the geneticist looking at the origins of his or her own discipline, the transition from eighteenth-century materialism to nineteenth-century developmentalism seems a retrograde step. Instead of looking for material links between parents and offspring, the embryologists were concerned only with the process by which the potential within the fertilized egg was unfolded. Their approach was either vitalistic or at least openly teleological; hardly a suitable prelude to the materialistic emphasis that would form the basis for modern genetics. As we shall see in the next chapter, Darwin's theory of natural selection introduced a disturbing factor that had the potential to threaten the developmental world view, and – more important for our present topic – reintroduced a genuine concern for the question of inheritance. Yet Darwin's impact was muted by the fact that he too shared a developmental model of reproduction, and it was not until the end of the century that a serious exploration of the hereditarian assumptions implicit in the selection theory could be undertaken. In the mean time, various non-Darwinian and non-Mendelian theories of evolution and heredity flourished under the aegis of developmentalism. This whole approach seems to bypass the issues that interest the modern geneticist, but it is precisely for this reason that we need to study it. Only by understanding the pre-Mendelian framework of nineteenth-century thought can we hope to appreciate why genetics emerged only as that century drew to a close.

On further reflection, even the geneticist should be able to recognize that the conceptual revolution of the early nineteenth century was not a dead end. Only by rejecting the rather crude materialism of the eighteenth century was it possible for biologists to escape the conceptual scheme in which the living organism is simply put together like a jigsaw puzzle. When materialism renewed its assault on the problem of sexual reproduction in the late nineteenth century, it had to search for a more sophisticated model which would allow a genuine process of development to take place under the control of material forces. If the structure of the whole organism was somehow encapsulated in the fertilized ovum, the process of unfolding would have to involve the translation into physical reality of information encoded within structures inherited from the parents. The early geneticists had little real interest in how that decoding took place, but their theoretical scheme would have been unthinkable without a recognition in principle that a developmental process intervened between the germinal material and the fully-formed organism.

The embryologists themselves paved the way for an exploration of how the process of inheritance might work, because their discoveries served as a channel by which the newly emerging cell theory could be applied to the phenomenon of reproduction. Von Baer's identification of the mammalian ovum was the first step in a recognition of how the process of fertilization takes place, while von Baer and many others traced the complex process by which the single cell of the fertilized ovum multiplies in the early stages of growth. Whatever the teleological assumptions of the early nineteenth-century embryologists, their work fed into the great development of cytology that created the empirical framework within which the concept of nuclear preformation – and hence the theory of the gene – would be formulated.

3

Evolution and Heredity

The Darwinian revolution had immediate implications for the study of heredity. Any theory of evolution implies a breakdown of the hypothetical guarantees introduced by earlier biologists to ensure that the basic form of the species is copied accurately from one generation to the next. New characters must appear and then be maintained within the species, so evolutionism necessarily focused attention on the relationship between variation and inheritance. Once Darwin's *Origin of Species* had triggered off the general conversion of the scientific world to evolutionism, the question of heredity would almost inevitably be drawn back onto the centre of the biological stage.

To understand the impact of evolutionism, though, we must make a distinction that is seldom recognized in orthodox historical accounts. There are many different theories about how evolution might work, all with their own implications for heredity. Most accounts of the Darwinian revolution – including those in the histories of genetics – concentrate almost exclusively on Darwin's mechanism of natural selection. This is the mechanism still accepted by modern evolutionists, and with hindsight it is easy to see how Darwin's theory contained the seeds of the concept of 'hard' heredity later enshrined in Mendelism. But Darwin's theory was not the only version of evolutionism available in the late nineteenth century. Many so-called 'Darwinists' made little real contribution to the research programme sketched out in the *Origin of Species*. Lamarck's theory of the inheritance of acquired characters was accepted in a subordinate role by Darwin himself, and was given even greater prominence by other evolutionists. By the end of the century, a fully fledged 'neo-Lamarckian' school had emerged in deliberate opposition to Darwinism, while other non-Darwinian ideas were also in circulation. Biologists investigating the problem of heredity thus had to pick their way through a maze of different evolutionary theories if they hoped to gain any inspiration from this source.

If we accept that non-Darwinian evolution theories played a major role in late-nineteenth-century biology, we must reassess the traditional interpretation of the Darwinian revolution as a prelude to the emergence of Mendelism. I have argued elsewhere (Bowler, 1988) that the most influential non-Darwinian theories were merely updated versions of the developmentalists' analogy between individual growth and the progress of life on earth. Darwin's essentially materialistic view of evolution was too radical for most of his contemporaries: the *Origin of Species* stimulated their conversion to evolutionism but did not convince them that natural selection was an adequate mechanism. Instead, they preferred to formulate alternative theories preserving the progressionist and teleological framework of developmentalism. Whatever the implications of Darwinism for the study of heredity, those implications could only be recognized by undermining the credibility of a host of rival theories. Even though our story will have to focus on the conceptual developments linking Darwinism to modern genetics, we cannot hope to understand why it took so long for those developments to bear fruit unless we recognize that they took place against a background in which a number of alternative theories helped to sustain a pre-Mendelian view of heredity.

Lamarckism and related non-Darwinian evolution theories worked best with a model of heredity in which the growing organism in effect 'remembers' the experiences of its parents and its more distant ancestors. The analogy with memory is a form of 'soft' heredity: what happens to the parents in the course of their lifetime must somehow be impressed on their offspring. Darwin's mechanism of natural selection could work with a model of 'hard' heredity: there was no need for the offspring to be imprinted with the effects of parental experiences, since variation was seen as being unconnected with the needs of the organism. Although Darwin's own theory of variation and inheritance was a developmental one, he accepted that most variation is essentially random. Individuals vary from the species' norm in all sorts of apparently meaningless ways, perhaps because their growth has been disturbed by changes in the external conditions. Only by chance will a few of these variants turn out to be useful, but since the individuals possessing favourable characteristics will do better in the struggle for existence, there will be a tendency for these characteristics to be transmitted more readily to future generations. Whatever the cause of variation (and we have completely rejected Darwin's own views on this topic), the selection theory requires that new characters, once formed, be transmitted as accurately as possible to future generations.

From our modern perspective, then, natural selection carries implicit within it the need for a theory of hard heredity. The orthodox story of the history of evolutionism (and of genetics) holds that the selection mechanism also requires heredity to be particulate rather than blending if it is to have a lasting effect in changing the form of a species. Genetics is thus portrayed as the missing piece in the Darwinian jigsaw puzzle, as though the selection theory created a conceptual vacuum that could only be filled by Mendel's laws. Darwin's failure to develop a modern approach to inheritance formed the only serious obstacle hindering his theory's acceptance, an obstacle that was removed as soon as Mendelism came on the scene. The purpose of this chapter is to undermine this traditional picture of Darwinism and Mendelism as consecutive steps along the straight and obvious path towards modern biology. We shall see that Darwinism could function quite effectively with a model of blending inheritance, and that the most powerful barrier to the theory's acceptance was not the lack of genetics, but the prevalence of developmental alternatives. The 'delay' in unpacking what we now see as the obvious hereditarian implications of Darwinism can be explained only by recognizing that the path from the selection theory to Mendelism was but one of many conflicting lines of investigation in late-nineteenth-century biology.

Non-Darwinian Evolutionism

On the historiographical model adopted in this chapter, Darwin's theory gained its initial reputation not because everyone accepted evolution by natural selection, but because the publication of so radical a mechanism acted as a catalyst to stimulate the supporters of developmentalism into action. Developmental models of evolution had already been suggested long before the *Origin of Species* was published, but had been greeted with indifference or hostility because as yet few naturalists saw the value of so openly challenging the traditional belief in the fixity of species. Darwin pioneered new lines of evidence that helped to convince everyone that evolution must occur, but rather than accept his combination of random variation and struggle, the developmentalists now turned in earnest to construct an evolutionary version of their own paradigm. To the extent that this paradigm dominated late-nineteenth-century evolutionism, we must accept that the aspects of Darwin's thinking that most impress modern biologists were not taken seriously by his contemporaries. The immediately post-Darwinian era did not exhibit a totally new world view, only an updated version of the old one. The story of how the more radical implications of Darwin's theory were eventually unpacked now becomes even more important, because we realize that to the majority of his contemporaries those implications were not as obvious as they appear by hindsight today.

Although the conventional image of the Darwinian revolution presents the *Origin of Species* as a bolt from the blue, there can be no doubt that the basic concept of transmutation had been discussed by radical thinkers throughout the early nineteenth century (for a survey of the history of evolutionism, see Bowler, 1984a). Even in Britain, where conventional piety still required all naturalists to pay at least lip-service to the idea of a divine Creator, the possibility that the development of life on earth had occurred without a series of miracles received wide publicity in the decades before Darwin published. The best evidence of this is the debate centred on Robert Chambers's anonymously published *Vestiges of the Natural History of Creation*, which appeared in 1844, fifteen years before the *Origin of Species* (Gillispie, 1951; Millhauser, 1959). Significantly, Chambers's theory was an amalgam of developmentalism and traditional British natural theology (Hodge, 1972). He used Meckel's law of parallelism to argue that the growth of the human embryo is the key to the progressive development of life on earth. For Chambers, evolution is nothing more than the gradual extension of the growth process further and further up the hierarchy of complexity. Each progressive step is achieved by the embryos of a whole species suddenly beginning to grow a little further before they mature. The theory takes for granted that the Creator has established a potential hierarchy of organic forms that will be gradually unfolded through the evolutionary process. Progress, not adaptation, was the key to the development of life. Chambers really had little use for the study of heredity because he took it for granted that once a new and higher character had manifested itself in a species, all subsequent generations would continue to mature at the same more advanced level.

Chambers's theory contains all the classic ingredients of a developmental view of evolution, with the teleological element fully displayed via the assumption that progress is the result of a divine plan. This latter claim was meant to deflect creationist attacks, but it did not prevent *Vestiges* from becoming the target for a barrage of theological and scientific criticism. Nevertheless, the book was widely read, and its progressionist image almost certainly helped to shape the climate of opinion into which Darwin's far more radical theory would be received. Professional biologists did not in general endorse Chambers's account, but many were now having second thoughts about creationism and were on the lookout for a new initiative that would allow science to address the problem of the origin of species. Darwin's book provided this initiative in 1859, offering a more sophisticated critique of creationism and an entirely new hypothesis about how evolution might work (discussed in detail below). An intense debate followed, but within a decade or so the majority of scientists – and the majority of educated laypersons – seem to have accepted the general idea of evolution.

But what *kind* of evolutionism did they accept? We know that the theory of natural selection encountered difficulties, but these have usually been dismissed as teething troubles caused by Darwin's failure to grapple adequately with the problem of heredity. Historians of biology are now coming to realize that the distrust of selectionism was much more broadly based. Even some of Darwin's staunchest defenders – including Thomas Henry Huxley – paid little attention to the selection mechanism in their own work. They were 'Darwinists' because they saw themselves as following Darwin's lead into an exploration of the general world of evolutionism, but they did not feel it necessary to adopt Darwin's explanation of how it worked. Huxley in particular was a morphologist who had little interest in the biogeographical evidence that Darwin used to support his vision of branching evolution. For him, it was enough to work out plausible lines of development linking the known fossils into evolutionary sequences (Desmond, 1982; Di Gregorio, 1984). Although suspicious of progressionism in both the early and the late phases of his career, in the 1860s Huxley succumbed to the popular belief that nature would tend inevitably to drive living things towards higher levels of complexity. For all his vociferous support – which earned him the nickname 'Darwin's bulldog' – there is a sense in which Huxley was only a pseudo-Darwinian rather than a true follower of Darwin's theoretical programme.

Although Huxley defended the *Origin of Species*, it was several years before he began to use the idea of evolution in his palaeontological work. His decision to begin using evolution as a guide to the fossil record was inspired by the writings of Germany's leading 'Darwinist', Ernst Haeckel. Where Darwin saw the imperfection of the fossil record as an obstacle to the evolutionists' hope of reconstructing the past history of the various living forms, Haeckel felt that evolution was worthless unless it could be used as the basis for a complete history of the organic world. His approach was frankly developmental: he adopted a Lamarckian explanation of adaptive evolution and fitted this into a progressionist vision in which the growth of the human embryo was the model for the general advance of life on earth. It was Haeckel, not Darwin, who popularized the recapitulation theory, the claim that ontogeny (Haeckel's own term for the growth of the individual organism) recapitulates phylogeny (the evolutionary history of the species). In popular works translated under titles such as *The History of Creation* and *The Evolution of Man*, Haeckel promoted an almost teleological view of evolution developing inexorably towards humankind as its goal.

Haeckel made no explicit appeal to a supernatural guiding force, so his progressionism was acceptable to those who saw evolutionism as a means of allowing science to function independently of religion. Materialism became morally acceptable through the supposition that nature had its own goals that would be achieved through an evolutionary process in which the human race is a key participant. This ostensibly naturalistic developmentalism inspired a whole generation of evolutionists both inside and outside science. Huxley was certainly impressed enough to begin his own search for evolutionary patterns in the fossil record. Although never an exponent of the recapitulation theory, he collaborated with other pseudo-Darwinists such as E. Ray Lankester who came even more closely under Haeckel's spell. In the 1870s, Lankester became the leading British exponent of the recapitulation theory, with its implicit assumption that there is a main line of evolutionary development leading towards humankind.

As Stephen Gould (1977) has shown, recapitulation was rendered more plausible by adopting a Lamarckian, rather than a Darwinian, explanation of the evolutionary process. If acquired characters are inherited, then variation can be seen as an *addition* to the growth of the individual organism. The acquired character eventually becomes the normal adult form, reached without any effort on the part of later generations. The original adult form has then become merely a staging post on the path to the new state of maturity. More generally, the whole sequence of ancestral adult forms is preserved in the growth pattern of later individuals, thus explaining recapitulation. Inheritance must be seen as a process of remembering the experiences of past generations, so they can be used as a guide to the development of the organisms that now make up the species. In an 1876 pamphlet, *Die Perigenesis der Plastidule*, Haeckel borrowed Ewald Hering's explicit analogy between heredity and memory, arguing that experiences are stored in wavelike motions within the 'plastidules' or basic units of living matter (Robinson, 1979, ch. 3). Hering's 1870 account of the memory analogy was translated by another prominent exponent of Lamarckism, Samuel Butler (1920, ch 6).

Although Haeckel attempted to explain memory/inheritance in apparently materialistic terms, his use of a concept associated with mental activity to illustrate a biological process was typical of his philosophy of 'monism'. This was a hybrid of idealism and materialism which left him free to attribute the properties of life to the basic units of matter. Small wonder, then, that Haeckel's progressionism went far beyond anything endorsed in the *Origin of Species*, for all that he called himself a 'Darwinian'. In fact, as we shall see, the selection theory does not require new characters to be seen as additions to growth, nor inheritance to be understood as a form of memory. Haeckel's popularization of the recapitulation theory thus ensured that Lamarckian concepts of inheritance would continue to obscure the hereditarian implications of Darwin's mechanism through into the later decades of the nineteenth century.

Lamarckism was popular even among thinkers who did not share Haeckel's commitment to recapitulation. The British philosopher Herbert Spencer was one of the most powerful advocates of evolutionism in both biology and the social sciences. Spencer was inspired by von Baer's model of development as a process of specialization, and thus avoided the linear progressionism implied by the recapitulation model. But he was also convinced that progress occurs through the effort and initiative of individual organisms. This has led to Spencer being branded a 'social Darwinist', although in fact he favoured a Lamarckian approach in which the effects of newly-acquired habits can be incorporated directly into the species' constitution. Spencer had advocated Lamarckism even before the *Origin of Species* was published, and his *Principles of Biology* of 1864 adopted a dual system of evolution based on the inheritance of acquired characters and the 'survival of the fittest'. To support this position he proposed a theory of inheritance in which 'physiological units' in the reproductive cells were endowed with the capacity to build themselves into the growing embryo. Since these units were subject to variation under the influence of changes in the parents' bodies, the inheritance of acquired characters was possible.

In later works, Spencer (1887, 1893) responded angrily to August Weismann's claim that acquired characters are not inherited. He became one of many critics of Weismann's 'neo-Darwinism', in which natural selection was the only conceivable mechanism of evolution. Spencer's reaction was typical of those evolutionists who felt that something more purposeful than the selection of random variations must be involved. The later nineteenth century saw an 'eclipse of Darwinism', as the majority of biologists turned openly against the selection theory (Bowler, 1983). Lamarckism and progressionism had always been popular, but the original form of Darwinism – what I have called pseudo-Darwinism – was flexible enough to allow support for other mechanisms in addition to natural selection. Weismann's dogmatism made such a compromise impossible, and in reaction to neo-Darwinism an increasing number of biologists began to regard themselves as opponents of the selection theory. Against neo-Darwinism stood neo-Lamarckism, in which the inheritance of acquired characters was hailed as the principal mechanism of evolution.

Deliberate opposition to Darwinism had emerged even earlier in some countries. The American school of neo-Lamarckism was founded in the 1870s by palaeontologists such as Edward Drinker Cope and Alpheus Hyatt. Their support for evolutionism had from the start been conditioned by the feeling that selection was not the real driving force. Inspired by Louis Agassiz's pre-evolutionary use of the analogy between individual growth and the history of life revealed in the fossil record, the Americans began from the assumption that evolution consists of a determinate sequence of additions to the process of individual growth (Gould, 1977; Bowler 1983, ch. 6). Independently of Haeckel they advocated both the recapitulation theory and the analogy between inheritance and memory. In Hyatt's case, enthusiasm for the parallel between evolution and the individual life-cycle led to the proposal that all lines of evolution are ultimately driven towards 'racial senility' as a prelude to extinction. Such views ensured that the Americans were unwilling to compromise with Darwinism. Their version of neo-Lamarckism anticipated by several decades the anti-Darwinian movements that would emerge throughout the world in the 1890s.

It would be easy to dismiss neo-Lamarckism as the last gasp of an exhausted conceptual tradition, soon to be swept away by the wave of experimental research into heredity that paved the way for Mendelism. Certainly, the palaeontologists who supported Lamarckism had little direct interest in the details of how inheritance is transmitted. But they were prominent biologists in their own field, and the indirect support they provided was enough to ensure that the inheritance of acquired characters could not be dismissed as without scientific foundation. Only towards the end of the century did the question of experimental support for Lamarckism become crucial. The new generation of biologists was determined to study only those problems that were accessible to experimental techniques. As we shall see below, it was this movement that helped to create the climate of hereditarian thought in which Lamarckism was tried and found wanting. Even then, however, it proved impossible to banish Lamarckism and associated developmental models from biology altogether. The more traditional disciplines such as palaeontology remained loyal to the old paradigm well into the twentieth century, ignoring the claims of the brash new science of genetics.

Darwin and Heredity

Since the theory of natural selection has become the basis of modern evolutionism, historians of biology have tended to assume that its discovery must rank as the most important step in the field's development. The classic notion of the Darwinian revolution is based on the assumption that Darwin established a new paradigm in biology. In thus destroying the traditional foundations of biological thought, he created a situation in which the emergence of Mendelian genetics would follow inevitably as a step towards the modern genetical theory of natural selection. We have now seen that the true story of nineteenth-century evolutionism is far more complex. Many naturalists at first preferred a developmental model of evolution which evaded the more radical implications of Darwin's theory. From their perspective, the *Origin of Species* was merely a catalyst that stimulated them to formulate their own version of evolutionism, not the foundation stone upon which a new and entirely materialistic biology should be built. Nevertheless, the development and publication of Darwin's theory remain of vital importance, partly because the appearance of so radical a mechanism did have a galvanizing effect upon other biologists, and partly because natural selection focused attention on certain issues that hindsight tells us were to become central to later developments. We cannot ignore the debate that raged around the selection mechanism – although we should be careful not to fall into the trap of thinking that selection was the only theory of evolution influencing the thinking of late-nineteenth-century biologists.

The orthodox picture of Darwin's achievement is based on the assumption that his acceptance of evolutionism was shaped by his early interest in biogeography. Convinced that evolution works by adapting populations to their local environment, and aware that migration constantly disperses sample populations to new environments, he conceived evolution as an irregularly-branching, open-ended process quite unlike the orderly progressive sequences of the developmentalists. Natural selection was purely a mechanism of adaptation: the random variation that Darwin saw as the raw material of evolutionary change would be quite incapable of driving the species in a predetermined direction. The struggle for existence merely picked out those individuals who by chance happened to vary in a favourable direction. They would breed more readily than unfavoured individuals and their adaptive characters would be passed on to the next generation. The theory requires that new characters must be inherited to a substantial degree by future generations, otherwise the advantage gained by favoured individuals in the struggle for existence would be dissipated. Heredity was thus a vital component of Darwin's theory, and according to the traditional historiography of evolutionism this is the one area where he failed to make a sufficiently radical break with the past. Darwin's theory of heredity – 'pangenesis' – assumed that inheritance is a process in which parental characters are blended smoothly together. But natural selection would not be able to work properly if blending were the case, since favoured characters would be diluted by intermixing in future generations. According to the interpretation proposed by Eiseley (1958), this problem was so severe that Darwin himself abandoned the selection theory for Lamarckism.

This interpretation is designed to enhance the view that Darwinism and Mendelism are the two important steps needed to create modern biology. Although it admits that the selection theory at first encountered problems, it keeps our attention fixed on the need for a better theory of heredity and encourages us to neglect the alternative developmental view of evolution. Mendel's work is seen as the logical answer to Darwin's problem, an inevitable continuation of the process that Darwin had initiated. This Whiggish interpretation is now crumbling in the face of historical research that has revealed the essentially non-Mendelian framework of nineteenth-century biology. We have already seen that many evolutionists were not agonizing over the weakness of Darwin's views on heredity, because they were too busy exploring non-Darwinian concepts for which the problem of blending was not an issue. Those who did accept the selection theory were able to do so because they did not see blending as the critical problem it appears to be through Mendelian eyes. Darwin himself is now known to have been deeply influenced by a developmental view of reproduction. It is meaningless to talk of him 'failing' to discover Mendel's laws, because his whole philosophy of evolutionism was conceived within a framework that would not have allowed him to appreciate the significance of Mendelian concepts. The crucial question that needs to be asked in assessing Darwin's role in the emergence of hereditarian thought is: how did a theory conceived in this way nevertheless contribute to a train of events that destroyed the developmental view of nature?

To answer this question we must distinguish the truly original aspects of Darwin's thought from those areas where he shared the prevailing views of his time. This is not the place for a detailed review of the immense literature on how Darwin was led to the concept of natural selection (for surveys see Bowler, 1984a, ch. 6;

Oldroyd, 1984; Kohn, 1985). Modern historians have endorsed and extended the traditional view that Darwin's most original insights came from his work in biogeography and his study of animal breeding. From biogeography – including his famous visit to the Galapagos Islands – he derived a conviction that evolution must be a branching process, not the ascent of a linear hierarchy. He then had to construct a hypothetical mechanism to explain *how* a population adapts to its local environment, and in search of clues he turned to the one area where significant changes within species were known to take place: the work of the animal breeders. Darwin soon became aware of the fact that breeders were somehow exploiting the natural variability that exists within any population. Having realized that the breeders worked by selecting out the 'best' individuals, he eventually came to see that the 'struggle for existence' caused by population pressure would perform a similar function in nature, constantly favouring the reproduction of those individuals that happened to possess a character better fitted to the environment.

One of the most important factors shaping Darwin's whole view of evolution was his acceptance of the fact that variation is essentially random. New characters do not appear preferentially in a single direction, least of all a direction that will be useful to the species. 'Random' variation does not mean *uncaused* variation, only that the cause operates in a way that does not predispose change to occur in a particular direction. Selection is necessary because there are many variants appearing within the population, and thus tending to cancel one another out. The handful of favourable characters must be picked out in a way that allows the individuals possessing those characters to breed more effectively. As long as their particular characters are transmitted to their offspring, selection will be able to increase the level at which the 'fitter' characters are represented in future generations of the species. The relationship between variation and heredity thus lies at the heart of Darwin's theory, and it is here that the latest research has shown how thoroughly his views were steeped in the non-Mendelian view of life.

Darwin's own explanation of variation and inheritance was the theory of pangenesis, published in the *Variation of Animals and Plants under Domestication* of 1868. A manuscript of the section on pangenesis exists from 1865, showing that immediately before publication Darwin added some material on the relationship between his theory and current thinking on the cellular structure of living matter (Olby, 1963). Gerald Geison (1969) argues that the final version of the theory was developed in the 1860s under the influence of Herbert Spencer's concept of 'physiological units'. But more recently Jonathan Hodge (1985) has shown that the origins of pangenesis go back to Darwin's earliest thoughts on the nature of the reproductive process, as recorded in his notebooks from the late 1830s. In Hodge's view, Darwin was a 'lifelong generation theorist': from the very beginning his approach to adaptive evolution was conditioned by his deep concern over the question of reproduction. The claim that pangenesis was an afterthought tacked on to the main edifice of evolution theory represents a distortion of Darwin's intellectual career promulgated originally in the *Life and Letters* (1887) by his son, Francis. This 'Franciscan' view of Darwin has been favoured by modern biologists precisely because it is intended to diminish the significance of the one aspect of his thinking that has turned out to be a blind alley.

According to Hodge's reading of the notebooks, the young Darwin was profoundly influenced by several factors normally dismissed as irrelevant by the orthodox or Franciscan view of his discovery. These include the theory of evolution suggested by his grandfather, Erasmus Darwin, in his *Zoonomia* (1794–6) and the Lamarckism of the radical anatomist Robert Grant, whom Darwin encountered during his brief studies at Edinburgh University (Desmond, 1984). Both focused on the problem of reproduction as the key to how new forms of life might appear, and thus form a bridge linking Darwin back to the eighteenth-century theories of generation proposed by Buffon and others. Darwin became convinced that the reproductive material from which the new embryo is formed has, in effect, been 'budded off' from the parents' bodies. Plants throw out buds as a means of growth, and if we accept the analogy with animal reproduction, the reproductive material in the parents' sexual organs is similarly a by-product of growth. The adult body produces a small-scale copy of itself built from material superfluous to its own growth. In asexual reproduction, the offspring is a mere copy of the parent, just like a plant grown from a shoot, but in sexual reproduction the parents' contributions are somehow blended together to form the germ from which their offspring will grow.

In the early 1840s (by which time he had already discovered the principle of natural selection) Darwin rethought his position on reproduction in the light of the physiology of growth outlined by the teleomechanist German biologist Johannes Müller (Lenoir, 1981, ch. 3; Sloan, 1986). Müller promoted an early version of the cell theory in which cellular structure was the foundation of all organic forms, but which allowed for the possibility that new cells might be formed within the organism from noncellular material. Darwin now began to argue that each tissue, perhaps each cell within the adult organism individually buds off minute particles (later called 'gemmules') which are passed through the body and concentrated in the reproductive organs. Fertilization required the blending together of gemmules representing the parts of both parents' bodies, and the character of the embryo would be determined by the particular combination of gemmules that came together to shape its growth. Because the gemmules were budded off from the parental tissues, Darwin believed that his theory unified the phenomena of sexual and asexual reproduction. The original production of the gametes which unite in fertilization was achieved by a process analogous to asexual generation in the parent organism. Although described in language typical of the early nineteenth century, Darwin's theory retained the central concept of the earlier mechanistic theories: the 'budding' model implied that the embryo is derived from particles produced by and somehow representing all the parts of the parents' bodies. However complex the actual process of development, the fertilized ovum contains information stored in the form of particles imprinted with the character of the organs from which they have been derived.

Darwin remained largely unmoved by developments taking place in cell theory during the 1850s. He believed that several spermatozoa were needed to fertilize the ovum, and that both sperm and ovum have been built from smaller particles supplied from all over the body. By the 1860s the majority of biologists had begun to suspect that cells cannot be built up in this way: a new cell can only be produced by the subdividing of an existing cell (see Chapter 4). In the final revision of his theory, Darwin tried to adapt it to these new developments, but he was committed to the view that particles smaller than cells are passed through the body to the reproductive organs. Pangenesis was widely condemned after its publication, several writers pointing out that Darwin seemed only to have revived ideas going back to Buffon, Bonnet and even Hippocrates. Many found the concept of myriads of particles traversing the body fundamentally implausible, while the consolidation of the new cell theory more or less guaranteed that pangenesis would have a very brief life.

Darwin himself nevertheless believed that his theory threw light on the bewilderingly complex problem of heredity and variation. He also took it for granted that pangenesis was a suitable foundation upon which to build his more general theory of evolution. The sheer number of gemmules ensured that in most cases the offspring would be a blend of characters derived from both parents – although in exceptional circumstances an uneven mixture might allow the character of one parent to predominate over the other. Gemmules might pass into the offspring without being developed, and could thus remain hidden while being passed on to future generations. This would explain the tendency for ancient characters occasionally to reappear as though from nowhere, the phenomenon known as 'reversion'. Pangenesis also allowed for the inheritance of acquired characters, as its modern critics like to emphasize. Because it followed the model of the eighteenth-century theories of generation, it inevitably allowed for changes in the parents' bodies to be reflected in the gemmules they produce, and hence in their offspring.

Useful characters acquired through exercise could thus be transmitted, but Darwin believed that most variation was not adaptive because it consisted of a random assortment of new characters. Much of his information came from the animal breeders he consulted during the years in which his theory was formulated. He did not think of variant characters as preserved within the population by fixed hereditary determinants. Instead he adopted an 'environmental' view of the origin of variation that had once been popular among the breeders (although many were now having second thoughts after finding that the character of races can be preserved in new environments). Darwin held that a change in the external environment interferes with the copying process by which the parental characters were imprinted on the gemmules, or with the process by which the characters were expressed in the growing embryo. The offspring thus tended to depart from the parental type, in effect leading to the production of new characters. On this view, domesticated species vary more than wild ones because they are forced to live in an unnatural environment. Hybrid variability – of special importance to a later generation of Mendelians – could be explained as a special case of environmental inducement, since the parents had been exposed to different conditions in the past.

Out of the range of random distortions thus introduced into the population, the breeders picked out those of value to mankind. Darwin argued that the struggle for existence would play a similar selective role in nature, providing a natural form of selection that would boost the frequency of reproduction of those individuals lucky enough to vary in the direction of increasing fitness. As long as individual peculiarities are to some extent transmitted to the offspring, the proportion of 'fitter' individuals in the next generation will thus be increased. Note that there is no implication here of variations occurring through the addition of new stages to growth. The new characters are produced by simply changing or distorting the growth process into a new pattern. There is thus no reason to expect that the old adult form will be preserved as an immature stage in the new pattern of growth. Darwin had little interest in recapitulation, although he did believe that the early stages of growth would offer clues to evolutionary relationships. Trivial new characters would be produced by disturbing only the later phases of growth, so the early embryos of related species might show an identity of form that would subsequently be lost as each developed towards its own specialized state of maturity.

The belief that pangenesis was somehow fatal to the plausibility of the selection theory has arisen because modern biologists have been anxious to display the superiority of the synthesis between Darwinism and genetics. Since heredity was the one field where Darwin's views had obviously led him off along the wrong track, it was easy to assume that this was the crucial source of doubt for his contemporaries. The excuse for promoting this interpretation of Darwin's situation was provided by the evident distress he felt when reading a review of the *Origin of Species* by the Scottish engineer Fleeming Jenkin (1867, reprinted in Hull, 1973). According to Eiseley's interpretation, Jenkin showed Darwin – and everyone else – that natural selection would be impotent if inheritance works by blending parental characters. In frustration, Darwin retreated more and more into Lamarckism, while evolutionary biology marked time until Mendelism came along to show that inheritance is particulate rather than blending.

Jenkin's attack centred on the claim that if inheritance requires blending, then the effect of favourable variation will be swamped because the handful of favoured variants will interbreed with the unchanged mass of the population. If an individual is born with a significantly new characteristic (a 'saltation' or 'sport of nature' in the language of the time), he or she can only breed with a normal partner and their offspring will only inherit one half of the new character. The next generation will only have one quarter, and so on – until after a few generations the dilution will have gone so far that no advantage is discernible. Even if the favoured 'sport' breeds more rapidly than other members of its species, the fact that it is but one individual within a normal population will ensure that its advantage is soon dissipated by interbreeding. Only if inheritance is particulate – if characters are inherited on an all-or-nothing basis with no dilution – will a favourable mutation be able to spread into the population.

The effect of this argument upon Darwin has been misunderstood through a failure to appreciate the strength of his commitment to evolutionary gradualism. Selection works by adding up minute changes over many generations, not by seizing upon the occasional sport or hopeful monster. Jenkin's review disturbed Darwin because he had adopted a very pessimistic view of the amount of variation available within a wild population (Bowler, 1974). Domesticated species live under artificial conditions which Darwin thought would tend to promote increased variation, but any change in the natural environment will be so slow that the amount of variation it can stimulate will be minimal. Darwin had, in fact, tended to write of favoured individuals (even those with only a minute advantage) as though they were very rare. In these circumstances, Jenkin's swamping argument would still apply, even though it had originally been intended to apply to large-scale variants or monstrosities.

In response to Jenkin's review, Alfred Russel Wallace wrote to Darwin insisting that natural variation is much more extensive than the swamping argument implies. *All* individuals vary from the hypothetical norm to some extent, and for any character (such as height in the human population) there will be a range of variation with most individuals concentrated around the mid-point of the range. If one side of the range – say taller than average – is favoured in a certain environment, then all of the individuals who are of above average height will benefit. Since by definition half the population is of above average height, there can be no question of a few favoured individuals being swamped by interbreeding with an unchanged mass. Darwin could evade Jenkin's argument by simply changing his model of variation within the population. If there are many individuals who share the favoured character to some slight degree, swamping cannot occur even if inheritance does involve blending. Jenkin had, in fact, admitted this, but had used a separate argument in an attempt to undermine the plausibility of selection acting on the normal range of variation. He claimed that selection would be able to change the population only within a well-defined limit, beyond which no further variation was possible. Darwin had already faced up to this argument by insisting that there was no good reason to suppose that such a limit could be maintained indefinitely.

The claim that natural selection was implausible when tied to a model of blending inheritance is thus inadmissible. Darwin had worked throughout with such a model, and Jenkin's critique served only to warn him of the dangers of postulating too small an amount of variation in wild populations. In the later part of the century, the biometrical school of Darwinism used statistical techniques to study the range of variation, and defended the selection theory on the basis of a model of heredity that still allowed for blending. The biometricians were, as we shall see in the next section, actively hostile to Mendelism precisely because it violated Darwin's emphasis on the continuity of evolution. Pangenesis certainly had its problems, but they did not derive from its link with blending inheritance. The selection theory also came under attack, but again the role of blending was minimal. The real barrier to widespread acceptance of the selection theory was the continued preference of most evolutionists for a developmental world view.

Darwin's failure successfully to dislodge developmentalism may in part have been a consequence of his own refusal to abandon the traditional belief that generation is the key to the stability or otherwise of species. For all his willingness to postulate undirected variation, he still saw new characters as the result of changes within the process of individual growth and reproduction. To the extent that an interest in generation remained fundamental to his theorizing, Darwin himself was a developmentalist. His theory introduced a major twist into the old paradigm, but it did not make a clean break with the past. The difficulties that Darwin experienced in convincing his contemporaries to abandon a developmental view of *evolution* stemmed in part from his own continued reliance on a developmental view of *reproduction*. Nevertheless, Darwin's new approach to evolution was one of the factors that helped to ensure that a number of later biologists would begin to explore the possibility of a break with developmentalism at the level of both evolution *and* heredity.

Ancestral Heredity

One of the most important lines of conceptual development linking Darwinism to the idea of 'hard' heredity centres on the work of Darwin's cousin, Francis Galton. From the start, Galton distrusted pangenesis and sought a new concept of heredity in which the germinal material is not manufactured by the parents, but is transmitted unchanged from one generation to the next. Rather than explore the physiological processes that might be responsible for this transmission, he developed statistical techniques designed to allow the study of heredity in large populations. Galton's 'law of ancestral heredity' derived an individual's inheritance from all its ancestral generations, not just from the parents. Although allowing for the blending of parental characters in the offspring, the law implied that the germinal material itself is *not* blended. It thus represents a half-way point towards the Mendelian notion of particulate heredity, making Galton a difficult thinker for modern geneticists to analyse. Asking whether he was or was not likely to have appreciated Mendel's work had he come across it before the 'rediscovery' is pointless, however. What *was* important about Galton's work was his general concept of heredity as a process independent of the productive capacities of the parent organisms. For a whole generation of biologists, this insight helped to clarify the field, displacing the more general notion of 'inheritance' used by Darwin and earlier naturalists.

At the same time, however, the effect of Galton's law was 'double edged' (De Marrais, 1974). For Galton himself, the law led to a break with Darwin on the mechanism of evolution. Convinced that 'ancestral heredity' would tend to blend out deviant characters, thus keeping the species true to the original type, Galton advocated a discontinuous model in which new characters could only be produced by saltations or 'hopeful monsters'. The swamping effect would not then apply because the new character would establish a new 'norm' for the species around which all future generations would tend to cluster. Because of this support for discontinuity in evolution, Galton was able to side with early Mendelians such as William Bateson, whose notion of particulate heredity had in part been derived from a belief in saltative evolution. Yet several of Galton's followers saw the law of ancestral heredity in an entirely different light. Led by Karl Pearson, the biometrical school used Galton's statistical techniques to support a continuous model of evolution based on the natural selection of minute variations. For Pearson, continuous evolution meant that inheritance could not be particulate, and the biometrical school vigorously opposed the introduction of Mendelism after 1900.

Although Pearson saw Galton's notion of hard heredity as a means of clarifying the role of natural selection, Galton's own work cannot be seen as a simple continuation of Darwinism. Like many of his contemporaries, Galton was swept up in the wave of opposition to the selection theory which characterized the last decades of the nineteenth century. Nevertheless, there are several aspects of his position which could not have been formulated without the stimulus of Darwinism. Despite his enthusiasm for saltative evolution, Galton's sense that the organism is shaped by the whole of its ancestry could only have been formulated within the general context of an evolutionary view of life's history. His statistical approach clearly exploited the Darwinian technique of visualizing evolution as a process taking place within a population, not just within individual acts of reproduction. While searching for the sources of Galton's new initiative, we must also bear in mind the extent to which he drew upon and extended the link between evolutionism and the study of heredity pioneered by Darwin.

Galton made a deliberate attempt to test the validity of pangenesis by breeding from pure-bred rabbits which had received blood transfusions from rabbits of a different colour (1871; see Robertson, 1979, pp. 20–1, 33–4). If the gemmules postulated by Darwin were circulated through the bloodstream, some would be transferred along with the blood and the rabbits' offspring would be expected to show a trace of the other breed's colouration. The results were negative, and although Darwin protested that he had not necessarily thought of the gemmules being transmitted in the blood, Galton was already convinced that pangenesis was flawed. In an earlier paper (1865) he had come out strongly against the inheritance of acquired characters and in favour of heredity as something more than a simple transmission of the parents' own characters. The constancy of human races – often over a wide range of geographical conditions – indicates that character is determined not by the parents (who may be influenced by local conditions) but by the constitution of the ancestral group established over many generations.

It seems evident that Galton's conversion to a belief in hard heredity was shaped by his social attitudes. (For general studies of his life see the autobiography, 1908; Pearson, 1914–30; Forrest, 1974; on ancestral heredity see Swinburne, 1965; Cowan, 1972a; on the social background see Cowan, 1972b, 1977; Mackenzie, 1982.) Robert Olby (1985, ch. 3) suggests that Galton's position was at first defined in opposition to the historian Henry Thomas Buckle's claim (1857) that the environment determines the course of a nation's destiny. Raymond Fancher (1983) has shown that early experiences in Africa convinced Galton not only of the inferiority of the black races, but also that their status as a distinct type of humanity was rigidly determined by heredity. The studies that culminated in his *Hereditary Genius* (1869) were intended to show that human mental characteristics are determined by heredity, not by education and upbringing. Galton thus did much to establish the hereditarian position in the 'nature versus nurture' controversy and became the founding father of the eugenics movement (see Chapter 8).

Galton's hereditarian social philosophy helped to define crucial aspects of his scientific thought, although it would be misleading to call him a 'social Darwinist'. His assumption that the human races have distinct characters preserved by heredity made him suspicious of Darwin's evolutionary gradualism. Contrary to popular belief, his support for a policy of social improvement based on limiting the reproduction of 'unfit' members of the white population was *not* based on an analogy with evolution by natural selection (although the concept of selection itself was, of course, involved). For Galton, each species and each racial group had its own typical character, and the effect of heredity was to preserve the type. Selection could improve the quality of a type, but could not lead to significant changes. Evolution could only occur through the appearance of a totally new type by saltation (from the Latin *saltus*, a leap, or what we would now call a macromutation). In his *Natural Inheritance*(1889) he used the analogy of a polygon rolling along a flat surface to illustrate the difference between the two types of variation. The polygon is relatively stable when resting on one face: minor disturbances may rock it about but it will always settle back into the original position. Only when a large disturbance pushes it so far that it rolls over onto the next face is any permanent change in the situation effected. The rocking about on a single face is equivalent to the normal variation responsible for individual differences, while tipping over onto a new face is analogous to a macromutation producing a permanent change in the species' character.

At the same time, Galton became convinced that individual character within a homogeneous population is determined rigidly by inheritance. A person's character cannot be significantly improved by a better upbringing, and any slight improvements that do occur will not be transmitted to the next generation. The law of ancestral heredity was designed to show how the past determines the present character of the population *and* to explain why inheritance is seldom a straightforward process. There is clearly a tendency for parental characteristics to be inherited, but everyone knows of exceptions to this, such as cases in which a clever father has stupid children. Galton argued that such anomalies occur precisely because the character of the individual is determined not by the parents alone, but by the whole line of ancestral heredity stretching back to the origin of the type. The law of ancestral heredity states that the individual's character is fixed by the following formula:

$$\tfrac{1}{4}p + \tfrac{1}{8}pp + \tfrac{1}{16}ppp \ \ldots\ldots$$

where p denotes the parents' contribution, pp the grandparents', and so on. Even distant generations transmit some of their characteristics directly to the new organism, and this may sometimes turn out to be decisive. The theory of ancestral fractions could thus explain the occasional cases of 'reversion' where an individual seemed to exhibit an ancient character that had apparently been eliminated from the breed. The Mendelians would reject this explanation, but were forced to invent a more complex mechanism to account for apparent cases of reversion.

Since the parental fraction is derived from the father and the mother, Galton's law allowed for blending inheritance. But as Olby (1985, pp. 53–7) shows, Galton saw the blending as a mixture of still distinct characters. The offspring of a black and a white animal might be grey, but the greyness would be composed of a mixture of black and white cells. It never occurred to Galton that the segregation of characters might affect the whole organism, and he thus came nowhere near anticipating Mendel's laws. Yet his belief that heredity represents a package transmitted intact from a whole series of ancestral generations provided a model that would be immensely useful for later workers in the field. He introduced the concept of the 'stirp' (from the Latin *stirps*, a stock or stem) to denote the sum total of germinal material transmitted to the individual. Here was the essence of the notion of hard heredity: a bundle of characteristics transmitted to the new organism *through* the parents, but representing a fixed inheritance not subject to any influences arising from changes in the parents' bodies.

Although Galton did not anticipate the Mendelian concept of segregation, his willingness to think of the effect heredity had upon the whole population allowed him to break with the traditional view of heredity and variation. Even Darwin had thought of these phenomena as two antagonistic forces, with variation constantly interfering with the tendency for parental characters to be copied in the offspring. Galton now saw that variation and heredity were merely different aspects of the same phenomenon: the preservation by heredity of a range of different factors existing within the population as a whole. The offspring sometimes differs from its parents because it has developed a character transmitted from a more distant ancestor, implying that characters can lie hidden for a number of generations without manifesting themselves. Variation occurs within the population because heredity transmits a variety of different characters, allowing them to be combined and recombined through sexual reproduction, sometimes with unexpected results. We can now see that variation within a population is maintained because of heredity: there is no need to postulate a force interfering with heredity, providing we think of variation as a function of the population, not of individual acts of generation. This transition to seeing variation as a feature of the whole population can be thought of as something lying implicit within the Darwinian selection theory, but obscured by Darwin's continued reliance on a generational model of sexual reproduction as the source of variation.

Galton certainly believed that the ancestral contributions were preserved in the form of material particles transmitted through the cells of the reproductive system. The units composing an individual's inherited stirp will multiply in the germ cells and compete among themselves to see which will contribute to the next generation. Unlike the later Mendelians, Galton thought that the material particles representing a particular character were capable of indefinite division, so that ever-smaller quantities could be transmitted over many generations. This assumption raised no problems because Galton deliberately turned his back on the physiological question of how the transmission takes place. Nor did he concern himself with how the germinal material controls the development of characters in the growing embryo. Since he was in any case interested in the behaviour of characteristics circulating within the whole population, he turned to statistics as a means of studying the effects of heredity at this level. He knew that in most cases the variability of a particular character within the population shows a 'normal' or 'Gaussian' distribution. This is the familiar bell-shaped curve indicating that the bulk of the population is clustered around the mean value for the character, while ever-smaller proportions are to be found towards the limits of variation in either direction. He believed that this distribution would be preserved over many generations if the character was controlled solely by heredity and not by environmental factors.

Galton argued that under normal circumstances the effect of the law of ancestral heredity would be to prevent variation from ever going too far from the norm. If an unusual combination of ancestral characters produced an individual at the extreme edge of the variation range, this individuals' offspring would always differ less markedly from the norm. Galton called this effect 'regression' and believed that it would guarantee the overall stability of the species or race until a saltation came along to establish an entirely new norm towards which future generations would tend to regress. Selection might produce a temporary distortion of the normal distribution, but it could not shift the mean and hence could have no permanent evolutionary effect. The law of ancestral heredity was thus held to support Galton's belief in discontinuous evolution, while never quite breaking through to the idea of discontinuous inheritance.

Galton's rejection of continuity in evolution separated him from the Darwinian camp, but his advocacy of a populational approach to the phenomenon of heredity and variation was clearly an extension of Darwin's insight that selection is a process that can only be seen to act within a population. In effect, he used the concept of hard heredity to clarify the principle of natural selection by showing that the mechanism could work independently of Darwin's own commitment to a developmental view of variation. Galton's decision to set aside the whole question of how reproduction and growth work at the physiological level allowed him to make a clean break with existing traditions in a way that Darwin could not. Evolution was uncoupled from the problem of generation, and in one stroke Galton undermined the whole complex of ideas that had upheld the developmental world view. Since individual organisms merely transmit ancestral germ plasm to future generations, there was no reason to suppose that inheritance involved the remembering of past experiences. Lamarckism, the recapitulation theory and the analogy between growth and evolution simply went out of the window as an inevitable by-product of the decision to treat variation as a phenomenon to be understood not as an extension of individual growth, but as the maintenance of hereditary factors within the population. Whatever the limitations he imposed on the effectiveness of natural selection, Galton's initiative in the field of heredity paved the way for the complete destruction of the developmental world view by challenging the link between generation and evolution that even Darwin had left intact.

Galton's statistical techniques became the chief research tool of the biometrical school. His disciple, Karl Pearson, took over both the populational approach to variation and Galton's rigid hereditarian view of society. According to Mackenzie (1982), Pearson's statistical innovations were designed to produce evidence in favour of hereditarian social policies. But in one crucial respect Pearson differed from his teacher: he remained true to Darwin's theory of gradual evolution through the natural selection of minute variations (Norton, 1973; Provine, 1971). Towards the end of the century he argued (1896, 1898) that the law of ancestral heredity would allow selection gradually to shift the mean towards which the population tends to regress, in effect producing a permanent change in the species' character just as Darwin had supposed. Experiments by Pearson's colleague W. F. R. Weldon (1894–5) were intended to show that selection had a slight but definite effect upon a wild population exposed to new conditions. In the depth of the 'eclipse of Darwinism', the biometrical school remained as one of the few sources of support for the theory of natural selection. Their work kept alive the link between Darwinism and the statistical study of variation and was to play a vital role in the subsequent creation of population genetics.

For the time being, however, Pearson's commitment to gradualism in evolution brought him into conflict with one of the most outspoken advocates of saltative transmutation, William Bateson. In the course of the 1890s, Bateson's morphological work had led him to reject the claim that evolution is a process of gradual adaptation. He advocated evolution by sudden saltations, and openly criticized the Darwinians. Immediately after the 'rediscovery' in 1900, Bateson took up Mendelism, seeing its discontinuous model of heredity as the perfect complement to his theory of discontinuous evolution. Inflamed by their previous hostilities, Pearson was almost inevitably led to reject Mendelism as incompatible with the observed range of continuous variation which exists in most populations. The resulting controversy was to have serious effects upon the future development of both genetics and evolution theory. Historians of genetics frequently dismiss the biometricians' attack on Mendelism as some form of mental aberration, the wilful turning of a blind eye to the obvious superiority of the particulate model. Yet the biometrical approach was itself an offshoot of Galton's more sophisticated view of heredity. However obvious the superiority of Mendel's laws may seem today, we must remember that much of the variation observed within large populations *is* of a continuous nature. Mendel's discontinuous characters may have held the key to a new and far more fruitful approach to heredity, but his laws had no immediately obvious application in the many cases where a species exhibits a continuous range of variability. Crossing tall and short varieties of peas will appear of little relevance to a biologist studying, for example, the distribution of height in the human population.

As we shall see, the apparent inconsistencies could certainly be ironed out. Mendelism is compatible with a continuous range of variation, but only when it is realized that the genetical structure of a wild population is far more complex than anything observed in an artificial breeding experiment. In the early years of Mendelism, the degree of mental flexibility that would have allowed such a reconciliation to come about was not available. Conceptual and methodological differences were exaggerated by professional rivalries which ensured that compromise positions would not be explored for some time. The biometricians may have been 'wrong' to reject Mendelism, but by the standards of modern evolutionism *neither* side in the debate had a clear picture of how heredity directs the flow of characters within a natural (as opposed to an artificial) population. To take a more positive line, it would be better to see Galton as the source of two important lines of theoretical development which would eventually be reconciled in modern population genetics. Biometry preserved the Darwinian insight that evolution must be seen as a change taking place within the whole population, while the Mendelians applied Galton's discontinuous model of evolution to the study of heredity. Both sides were committed to the concept of hard heredity, whether applied to whole populations or discontinuous variants. As it turned out, clarification of the laws of inheritance would come first through the discontinuous approach, but the biometricians' statistical techniques would be needed eventually if the new genetics was to be applied to large populations and to evolution theory.

The eventual reconciliation of the biometrical and Mendelian approaches suggests that the orthodox historiography in which Darwinism was obviously 'incomplete' without Mendelism is in need of serious revision. If we must apply hindsight to the situation, what Darwinism needed was the concept of hard (but not necessarily particulate) heredity. Pearson expounded a detailed and plausible theory of natural selection which was not based on particulate inheritance, and refused even to accept that Darwinism was compatible with Mendel's laws. The greatest threat to Darwinism in the late nineteenth century had come from the alternative developmental view of evolution, which retained soft heredity as an essential component of Lamarckism. But by retaining the view that heredity and variation should be understood in terms of individual acts of generation, Darwin left one of the major foundations of the developmental viewpoint intact. Galton's concept of hard heredity reinforced Darwin's claim that variation is essentially random, because it completed the elimination of both Lamarckism and the analogy between growth and evolution. Whatever the process by which new characters were introduced into the population, it was not directed by the growth patterns of individual organisms. The claim that parents do not manufacture germinal material in their own likeness threatened the whole traditional edifice of biological thought. Whether or not we regard Galton's concept as lying implicit within the selection theory, his initiative certainly had the potential to complete the destruction of the developmental world view that Darwin had begun.

4

The Cell in Development and Heredity

The title of this chapter is borrowed from E. B. Wilson's classic survey of cell theory (significantly, the first edition of 1896 had been entitled *The Cell in Development and Inheritance* – the switch to 'heredity' illustrates the growing refinement of terminology at the turn of the century). Wilson's book reviewed the contributions made by the one area of biology upon which Galton had deliberately turned his back. Indeed, it could be argued that Galton's statistical approach was all the more radical because it uncoupled the question of heredity from the detailed study of the reproductive process. The development of cytology (the study of cells) in the late nineteenth century certainly paved the way for the classical genetics of the early twentieth. By the 1890s, August Weismann had synthesized developments in evolution theory and cytology to provide an independent line of argument for hard heredity. It is Weismann, not Galton, who immediately springs to mind as the originator of the claim that germinal material is transmitted through, but remains independent of, the parents' bodies. The associated belief that characters are defined by material units preserved but not manufactured in the reproductive system played a vital role in destroying the developmental view of generation and evolution. And yet the very fact that cytology allowed access to the problems of reproduction and growth – the central themes of the old theories of generation – meant that clarification of the concept of heredity via this route would be a slow and tortuous process.

Darwin himself only belatedly began to take note of the new cell theory that had emerged by the 1860s. In fact, the theory now created major problems for pangenesis because it was accepted that new cells (including the egg and sperm) cannot be built up from noncellular particles such as gemmules. Cells can only be produced by the subdivision of existing cells, and the central problem of nineteenth-century cytology was: how can this fact be used to interpret the phenomena of reproduction and growth? According to the orthodox historiography, Weismann sketched in the correct answer to this question when he began to argue that development (and hence the characters of the adult organism) is controlled by and in effect predetermined within the chromosomes of the cell nucleus, and that the chromosomes represent germinal material perpetuated through the division of the reproductive cells. The germ plasm controls the development of the organism and is retained by the adult organism within its reproductive cells, where it is isolated from any influences stemming from changes to the adult body. By insisting that there can be no feedback from the adult body to the germ plasm transmitted to future generations, Weismann independently pioneered the concept of hard heredity.

In principle, Weismann's theory implied that individual development should not be used as a model for evolution, because there is no way in which the growing organism can influence the production of new characters within the germ plasm. Yet modern historical research suggests that Weismann's ideas arose from within the developmental view of heredity and variation to which Darwin himself had contributed. For Weismann and most of his contemporaries, as for Darwin, reproduction and individual development were still aspects of a single, integrated biological process. Weismann's views on 'nuclear preformation' were controversial not only because he repudiated Lamarckism, but also because many embryologists still disputed the claim that development is controlled solely by predispositions built into the fertilized ovum. Those who saw development as a process requiring continuous interaction with the embryonic environment were particularly unlikely to endorse any challenge to the traditional view that the appearance of an organism's characters cannot be understood without taking the developmental process into account.

It is significant that Weismann's theory should have been described as a revival of preformationism. The source of the analogy is obvious enough, but should not blind us to the scope of the theoretical developments separating the old and the new versions of this concept. In the old preformation theory, the growth of a new organism was nothing more than the expansion of pre-existing parts. The structure was already there as a miniature within the fertilized ovum, the tissues needing only to expand in size to generate the adult organism. 'Nuclear preformation' in the late nineteenth century implied that the *potential* character of the organism was predetermined, but the theory could no longer rely on growth (i.e. simple expansion) to explain the development of the embryo. Of course tissues must expand in bulk during growth, requiring the duplication of cells with identical characters. But the structure of the embryo is developed gradually from a single cell, the fertilized ovum, by a differentiation that generates a bewildering yet orderly array of cells performing different functions. One of the most crucial questions faced by the new preformationists was: how does this development take place? They needed to explain not only growth (duplication) but also development or differentiation, which required them to account for the parcelling out and actualization of hereditary information within the tissues of the developing embryo.

Even the preformationists thus believed that it was essential for their theory to explain not only how information is transmitted from parent to offspring, but also how that information is expressed in the growing organism. The fact that transmission and development were still not regarded as distinct areas of biology illustrates the continued influence of the developmental model even upon biologists who were trying out concepts that would eventually undermine that model. Modern genetics emerged when the study of development (embryology) was separated off from the study of how characters are transmitted. It was this event which finally broke up the developmental theories of evolution and made it possible for the significance of Mendelian-style experiments to be appreciated. Historians are increasingly beginning to suspect that Mendel's laws remained unrecognized precisely because no one in the period before 1900 was prepared to consider a study of heredity that was so obviously divorced from the problem of growth (Sandler and Sandler, 1985; Horder, Witkowski and Wylie, 1986). Galton was able to clarify the concept of heredity because his population studies broke with the developmental tradition and could thus influence those biologists who were still struggling to untangle the physiological processes involved in reproduction and growth.

Although the cytologists were slow to recognize what we now regard as a fundamental distinction, their studies helped to create a revolution in biological methodology. Haeckel's developmental evolutionism was by its very nature highly speculative. It used morphology (the study of organic form via comparative anatomy and embryology) to reconstruct the history of life on earth by proposing hypothetical phylogenies to fill in the gaps in the fossil record. Cytology focused attention on to the study of reproduction and growth for its own sake, and by the end of the century many biologists interested in the problems of variation, heredity and evolution had begun to lose patience with the element of speculation that entered into any detailed exploration of the analogy between growth and evolution. For them, experiment was the key that would unlock a new level of biological understanding. If the hypothetical lines of ancestry postulated by Haeckel and his followers could not be tested by hard evidence, then this aspect of evolutionism should be left on one side in favour of direct studies of how variation and inheritance actually worked. A new generation of embryologists studied growth not as a clue to evolutionary relationships, but in the hope of discovering the physiological processes that actually built up the structure of the developing organism. In Britain, William Bateson turned in disgust from the ancestry of the vertebrates to the direct study of discontinuous variations, a move which paved the way for his eventual conversion to Mendelism. By themselves, such moves did not break down the developmental view of inheritance, but they helped to create new biological disciplines which sidestepped the analogy between growth and evolution that lay at the heart of the Haeckelian tradition.

At first the new disciplines were pioneered by European biologists. Cell theory and experimental embryology were largely the products of German science, with British and American biologists acutely aware of their own provinciality. By the end of the century, however, American biologists in particular were beginning to gain a new level of confidence as their scientific community grew in power and influence. Genetics was to become one of the first areas of biology where Americans took the initiative ahead of their European counterparts. Garland Allen (1978) wrote of a 'revolt against morphology' among American scientists at the turn of the century – a deliberate turning away from the old natural history tradition towards the new experimental biology. The initial formulation of this thesis has been challenged (Maienschein, 1981; Benson, 1981; Rainger, 1981) but Allen has reformulated his position (1981) in a way that retains his belief in a fundamental methodological revolution in American biology at the end of the nineteenth century. Whether the transition was a genuine revolution or merely a rapid shift of priorities, the period certainly saw major developments that were to pave the way for new science such as genetics in the twentieth century.

The Mechanism of Development

Although the development of cytology and the experimental technique helped to pave the way towards modern genetics, their initial effect did not include a fragmentation of the traditional concept of generation. Studies of fertilization and studies of growth were conducted as parts of a general assault on the question of reproduction. From the perspective of a modern geneticist, the most important outcome of this process was the eventual synthesis of breeding studies with experimental work on the process of fertilization to form the basis of classical genetics. But to many late nineteenth-century biologists it looked as though embryology would be the main beneficiary of the new techniques. Instead of playing second fiddle to evolutionism via the recapitulation theory, embryology would enjoy a new lease of life through its investigation of the causes underlying the growth of the embryo from the fertilized ovum. One of the greatest American geneticists, Thomas Hunt Morgan, began his career as an experimental embryologist and only abandoned this for the study of heredity some years after the emergence of Mendelism. To some extent, at least, the foundations of modern genetics were based on the growing frustration of embryologists such as Morgan who found that – whatever their initial expectations – the experimental technique was not capable of providing them with a rapid solution to the problem of growth.

From the time of von Baer onwards, it had been obvious that the embryo grew from a single cell. Haeckel and the other evolutionary morphologists knew that the early stages of development had to be understood in terms of the behaviour of the cells formed by the subdivision or cleavage of the fertilized ovum. Although committed to a careful description of this process, their interest lay to a large extent in the possibility of finding clues to the early development of life on earth. By the 1880s, though, the attractions of the recapi-tulation theory had begun to pall, allowing embryology once more to emerge as a field of study in its own right. It was the German embryologist Wilhelm Roux who found the science of *Entwicke-lungsmechanik* with the aim of providing a mechanistic explanation of how the embryo develops (Churchill, 1973; Allen, 1978, ch. 3; Maienschein, 1986). Originally a student of Haeckel, Roux rejected the search for phylogenies but took over his teacher's ostensibly mechanistic philosophy. The goal of embryology was to show how the structure of the growing organism is formed, and the tool it would use to perform this task was not observation but direct experimental interference with the growing embryo.

In a classic paper of 1888 (partially translated in Willier and Oppenheimer, 1964, pp. 2–37) Roux described an experiment which, he claimed, gave evidence that the development of the embryo is predetermined by material factors already present in the fertilized ovum. He used a needle to destroy one of the blastomeres (cells) formed after the first cleavage of a fertilized frog's egg. The embryo's subsequent growth was imperfect, leading Roux to argue that development must be predetermined by the material contents of the egg. The developing organism is not a flexible system capable of responding to challenges imposed on it from without. Roux also advocated a 'mosaic' theory of growth in which the process of cell division in the embryo partitioned out the germinal material to the daughter cells. Each cell receives only a portion of the organism's original germinal material, enough to control its subsequent development into a particular part of the body. August Weismann's theory of the germ plasm (discussed below) followed a similar model: the whole structure of the organism is predetermined by material elements located in the chromosomes in the nucleus of the fertilized ovum, and these elements are gradually parcelled out among the cells as they divide to form the tissues of the growing organism. Weismann's apparently modern concept of hard heredity was thus linked to a theory of growth which eventually had to be abandoned as cytologists showed that each cell retains the total genetic inheritance of the organism. Explaining how the cells are 'turned on' to different functions as the embryo grows was to prove beyond the capacity of the early experimental embryologists.

Roux and Weismann's concept of 'nuclear preformation' assumed that the germ plasm contains elements responsible for each character to appear in the growing organism. But many biologists were not persuaded that the embryo develops merely by unpacking characters already predetermined by inheritance. Lamarckians in particular were inclined to assume that the growing organism can respond in a positive way to challenges imposed by the environment. Some embryological experiments seemed to confirm the existence of such a flexibility, especially the work of Hans Driesch. In an experiment reported in 1892 (translated in Willier and Oppenheimer, 1964, pp. 38–50) Driesch took a developing sea urchin embryo and shook it so severely that the constituent cells became separated. According to the Roux–Weismann theory, the individual cells should not have been capable of proper development since they were not programmed to deal with such a drastic change in their situation. Yet they did, in fact, grow into small but complete sea urchins. Driesch concluded that the organism's growth potential is not parcelled out among the dividing cells; each somehow retains the capacity to generate the whole organism. He also began to argue that growth is not totally predetermined by material elements built into the fertilized egg. Driesch proposed that growth is a flexible process shaped to a significant extent by the environment in which development takes place. The growing organism can adapt to changes imposed on it from without, and Driesch was inclined to accept the Lamarckian view that such adaptations might be transmitted to future generations. He favoured a holistic view of life and later adopted an openly vitalistic position in which a non-physical 'entelechy' supervised the creation of new organic structures.

Driesch's vitalism was not the only line of opposition to the idea of nuclear preformation. In 1894 Oscar Hertwig related the debate over embryological development to the old distinction between preformation and epigenesis (translation 1896). He argued that the Roux–Weismann theory amounted to little more than a return to preformationism. The elements in the germ plasm might not comprise a miniature organism, but the theory did postulate material particles that in some way contained the potential to generate each part of the new organism's body. In the absence of a satisfactory explanation of how the potential of the germ plasm is translated into reality, Weismann's theory evaded the real problem of growth just as the earlier theory of pre-existing germs had done. Hertwig preferred a theory of epigenesis in which there was a genuine process of development because the structure of the organism was not completely predetermined within the germ plasm. Like Driesch he believed that the growing embryo is a system with a certain amount of built-in flexibility, allowing it to respond to changes in its environment. The parts of the growing embryo are co-ordinated in a way that allows them to interact in an apparently purposeful manner. This would not be possible if each cell were programmed rigidly in advance by its inherited germ plasm.

In America, E. B. Wilson's *The Cell in Development and Inheritance* (1896) complained that the new debate of preformation versus epigenesis was as sterile as the old one. By the first decade of the new century, however, Wilson had become convinced that some characters are predetermined by the material structure of the chromosomes. His work helped to convince many other biologists that the cell nucleus was indeed the key to heredity. Yet some Americans continued to resist this claim (Maienschein, 1987). E. G. Conklin ignored the nucleus and favoured a view in which the cytoplasm (the extra-nuclear material of the cell) plays an important role in allowing the growing organism to interact with its environment. Conklin vigorously resisted any effort to separate the study of inheritance from the problem of growth. Even T. H. Morgan began his career as an experimental embryologist with a preference for epigenesis over preformation (Allen, 1978). In 1905 he still opposed Wilson's views on the importance of chromosomes and rejected Mendelism as an unwarranted speculation.

A particularly important problem centred on the question of sex determination (Maienschein, 1984). As the most fundamental character difference between organisms of the same species and the foundation of the reproductive process, biologists naturally sought to explain how the sex of the individual was determined. Was the sex of the embryo predetermined in the fertilized ovum, or was it decided by the conditions under which development took place? The Mendelians of the early twentieth century would ultimately decide that sex was predetermined by the inheritance of particular chromosomal factors – indeed, sex determination was to play a key role in the synthesis of Mendelism with the chromosome theory of heredity. Yet in the 1890s the situation was still unresolved, with many biologists preferring an epigenetic view in which the environment played the vital role. They held that an organism's sex was not predetermined by the inheritance of a material factor; instead each embryo started with the potential to be either male or female depending upon the environment in which it developed. The whole notion of predetermined character-units was unthinkable within this epigenetic approach, which thus stood opposed to anything like a Mendelian or a chromosomal theory of inheritance. Yet the epigenetic position cannot be dismissed as a complete fallacy. Although sex determination later became a key factor in convincing many biologists that characters *are* linked to chromosomes, this is true only for certain types of organisms (including, of course, those favoured as experimental subjects by the geneticists). In some reptiles, for instance, sex is decided by the environment, particularly the temperature at which the egg is incubated. There was thus no way of establishing an unambiguous resolution of the debate by observation or experiment. Only after years of debate did attention begin to focus on those cases in which sex is controlled by heredity, allowing this to emerge as a key example of Mendelian inheritance. In the early years of the twentieth century a consensus was established in which nuclear predetermination was held to be the source of sex determination in most animals. It was this issue which led Wilson to stress the role of the chromosomes and which ultimately turned Morgan from an embryologist into the founding father of classical genetics.

The controversies over development formed a vital component of the pre-Mendelian view of reproduction. Before 1900 the majority of biologists still saw development as an integral part of the reproductive process and were unwilling to concede that one could study the inheritance of characters without worrying about how the characters were produced in the growing organism. Experimental embryology, not breeding studies, seemed to offer the most exciting avenue leading towards a better understanding of reproduction. Some biologists still did not believe that the nucleus of the fertilized egg was programmed with all the information necessary to generate a complete new organism. Even when not explicitly linked to Lamarckism, these attitudes reflect the continued influence of the developmental tradition and formed a barrier against the emergence of a separate study of inheritance. When we turn to look at the origin of classical genetics (see Chapter 7) we shall have to investigate how some embryologists abandoned the hope of providing a complete explanation of how the organism develops and began to concentrate on the far more tractable problem of how characters are transmitted from parent to offspring.

The Germ Plasm

The biologists who favoured an epigenetic view of development inevitably diminished the role of 'heredity' (in the modern sense of the term). They believed that the environment created in the egg or womb was at least as important as the transmission of genetic information by the cell nucleus when it came to determining the character of the growing embryo. The rival preformationists were necessarily committed to something more in line with Galton's view of heredity. Some kind of material structure must be transmitted from parent to offspring, carrying programmed within it the information necessary for the construction of the new organism. The question was: what is the origin of this material structure, and how is it conveyed to the nucleus of the fertilized egg? Weismann's theory of the germ plasm was one of the most innovative (and controversial) attempts to solve this problem. In its own way it paralleled Galton's concept of hard heredity and thus helped to create the framework within which early twentieth-century biologists would try to understand the laws of inheritance derived from Mendel's work.

The key step implied by Galton and Weismann's ideas is the destruction of the traditional belief (still enshrined in Darwin's pangenesis) that the parents' bodies manufacture or bud off particles responsible for transmitting their characters to the next generation. Weismann argued that the new cytological discoveries eliminated this possibility and required one to accept instead his claim that characters are determined by material structures inherited by the parents from earlier generations and transmitted unchanged (except for the mixing required by sexual reproduction) to their offspring. In principle this theory undermines the foundations of the old developmental tradition: growth is irrelevant to variation, heredity and evolution because what happens in growth cannot affect the transmission of germinal elements or the creation of new ones. The recapitulation theory and the analogy between memory and inheritance lose their credibility because evolution can only take place through the production of new characters by some process operating purely within the germ plasm, not within the body as a whole. Yet modern historical research shows that Weismann's position at the intersection of evolution theory and cytology led to his theory being constructed within the framework of developmentalism. Although responsible for an idea that would help to destroy the link between generation and evolution, Weismann himself was unable to make the clean break with the past that would lay the foundations of Mendelian genetics.

The concept of a material substance responsible for transmitting the information necessary to build a new organism figured prominently in Carl von Nägeli's *Mechanische-physiologische Theorie der Abstammungslehre* of 1884 (conclusion translated 1898; see Robinson, 1979, ch. 6). Nägeli has had a bad press from historians of genetics because he was the one well-known scientist with whom Mendel was in regular contact. It has thus been assumed that Nägeli's unwillingness to take Mendel's laws seriously was a key factor blocking the path to more general acceptance. Instead of trying to incorporate the laws into his own theory, Nägeli ensured that Mendel's future work would be sidetracked by encouraging him to study a singularly difficult subject for genetical research, the hybrids of the genus *Hierarcium*. In fact, we shall see that the relationship between the two biologists has been misunderstood through a failure to appreciate that Mendel himself might have had goals that do not coincide with those of the modern geneticist (see Chapter 5). Nägeli is important not because of his 'failure' to appreciate the validity of Mendel's laws, but because his own theory helped draw attention to a very different idea that would eventually be incorporated into classical genetics: the concept of a material substance responsible for encoding the information transmitted by inheritance. For Nägeli, this substance – the 'idioplasm' – could be understood as consisting of *Anlagen*, a German term denoting an outline or rudiment, each *Anlage* being responsible for a particular character of the developing organism. The *Anlagen* were not preformed miniatures but predispositions that would shape the growing embryo in a particular way. The predisposition was somehow built into the material substance of the idioplasm, which Nägeli saw as being made up of units called 'micelles' with a crystalline structure.

Although neither Nägeli nor anyone in the next generation of biologists could propose a satisfactory explanation of how the potentiality of the *Anlangen* was stored and then translated into reality, the concept itself was vital to the foundation of an independent science of heredity. The 'gene' of classical genetics was a particle with precisely those unexplained properties postulated by Nägeli. Yet outside this one concept, Nägeli's theory did not contribute to future developments. In evolution theory he maintained that a 'perfecting principle' resided in the idioplasm, capable of affecting individual variation in a way that would drive the species steadily towards higher levels of organization. He also believed that the idioplasm constituted a network spread through the whole body of the parent organism. This meant that acquired characters might – if they persisted over many generations – be able to influence the idioplasm and thus become hereditary. Despite his allegedly mechanistic philosophy, Nägeli thus played only an indirect role in the creation of a materialistic concept of the hereditary substance.

The claim that the idioplasm was spread throughout the whole organism was treated with much scepticism because it seemed to violate principles already established by cytological research. Nägeli had virtually ignored the cell theory, and his concept of a hereditary substance would only become influential when other biologists, especially August Weismann, fused it with the latest discoveries and ideas in this area. In its original form, the cell theory proposed by Matthias Schleiden and Theodor Schwann had stressed the importance of the cellular structure of living matter, but had left open the possibility that cells could be built up from noncellular material in the body. By the 1850s many biologists had become convinced that cells cannot be formed in this way: new cells can only appear by the subdivision of existing cells. In the dictum popularized by the cellular pathologist Rudolph Virchow, *Omnis cellula e cellula* (all cells come from cells). This new interpretation undermined not only theories such as pangenesis, but also Nägeli's hypothesis of an idioplasm which linked and interpenetrated all the cells of the body.

By the 1870s, improved microscopes were available, and new techniques for staining cells to reveal their internal structure were beginning to focus attention on the nucleus (for details of developments in this area see Coleman, 1965; Robinson, 1979; Farley, 1982; Mayr, 1982, part 3). It would be a mistake to assume, however, that by themselves the new microscopic discoveries allowed biologists to solve the problem of heredity (Baxter and Farley, 1979). It was only with the conceptual innovation of the Mendelian-chromosome theory in the early twentieth century that a consensus began to emerge over the interpretation of what was going on within the cell during the reproductive process. Up to this point there was widespread debate over the meaning of the processes revealed by the microscope. Even when biologists agreed over what they were observing, they disagreed over the significance of the structures and processes that had been revealed. There was no universally recognized foundation in cytology that would lead to the acceptance of Mendel's laws. Instead, it was the growing popularity of the explanatory system which became the basis of classical genetics that at last allowed biologists to agree over the interpretation of their observations.

The nucleus was of particular importance to those biologists who favoured the view that a material substance is responsible for transmitting characters from parent to offspring. It was observed that at the time of cell division, small rod-like bodies appeared in the nucleus. These were eventually called 'chromosomes' because they absorbed the coloured stains used to highlight cell structure for the microscope. In 1879 Walther Flemming described the longitudinal splitting of the chromosomes during mitosis, or normal cell division. It was this process that would eventually lead the exponents of nuclear preformation to argue that the process is designed to produce exact copies of the original chromosomes in the daughter cell. When applied to the process of embryonic differentiation, this would disaprove the mosaic theory in which the original chromosomal material was thought to be parcelled out to the various parts of the developing body.

Other discoveries began to throw light on the process of fertilization. In 1875 Oscar Hertwig observed the sperm entering the ovum at the moment of fertilization. He argued that the fusion of the two nuclei was the object of fertilization and effectively eliminated the old belief (still accepted by Darwin) that several sperms were needed to fertilize a single egg. In 1883 the Belgian cytologist Edouard van Beneden reported studies of fertilization in the horse threadworm, showing that during meiosis (the quadruple cell-division which generates the egg or sperm) the gametes receive only half the chromosomes of the normal cell. Meiosis was thus known as 'reduction division', and it seemed evident that fertilization was necessary to restore the full complement of chromosomes by bringing together the halved contributions from both parents.

Weismann was one of the biologists who used van Beneden's observations to clarify his thinking on the process of fertilization. He believed that the material structures responsible for coding the units of hereditary information, the 'ids' as he called them, are strung along the length of the chromosomes. The chromosomes were permanent features of the cell, but in the case of reduction division half the parent's ids were lost, to be made up in the offspring by half of the other parent's units supplied at fertilization. Since the ids behaved as discrete units, there was no blending: the units are moved around through reduction division and fertilization but retain their individuality throughout the various combinations. In this respect Weismann disagreed profoundly with Oscar Hertwig, who maintained that the chromosomes are *not* permanent features of the cell (they are invisible when the cell is 'resting' between divisions). Hertwig's position seems to have been linked to his preference for a model in which the germinal materials fuse completely to give a blending of characters after fertilization (Churchill, 1970).

The mature version of Weismann's theory of the germ plasm (translation 1893a) aimed at a synthesis of the new cytological discoveries with his own conceptual model. This model had begun to mature in the course of studies Weismann undertook into various aspects of evolution theory, beginning with the development of colouration in insects (translation 1882). Although already a 'Darwinist', he did not at this time rule out the possibility of the inheritance of acquired characters or of directed variation. Stephen Gould (1977, pp. 102–9) has shown that these early studies depended on the logic of the recapitulation theory for their interpretation of the evidence. More recently Frederick Churchill (1986) has described how recapitulationism lay at the heart of Weismann's thinking during his crucial microscopical observations of Hydromedusae in the period 1876–83. It was his analysis of these observations that led Weismann to formulate his distinction between the germ plasm and the soma (the structure of the body as a whole). Although the concept of an independent germ plasm would ultimately destroy the plausibility of recapitulationism, Weismann's theory had its origins in studies which were a direct continuation of the Darwinian link between generation and evolution, conducted firmly within the developmental tradition.

On the strength of these studies, Weismann concluded that single-celled, asexually reproducing organisms are potentially immortal, since they multiply by simply dividing themselves. All life had once been like this, but when multicellular organisms at last evolved, they did so by a process of specialization which produced a rigid distinction between the reproductive cells (the germ plasm) and the cells from which the rest of the body is constructed (the soma). The germ plasm retained the potential immortality of the single-celled ancestors, its cells dividing within the reproductive system and passing their characters on without change to future generations. The somatic cells are derived from information supplied by the germ plasm, but they are sterile once the full differentiation of the body is complete. The organism as a whole is thus subject to death, only the germ plasm surviving if it has reproduced successfully. Once the somatoplasm has been constructed, it is unable to influence the germinal material for which it is the host. The organism can transmit a proportion of its inherited germinal material to its offspring, but it cannot affect the characters thus perpetuated because the reproduction of germinal cells takes place entirely separately from the rest of the body. In Weismann's theory there was a strictly one-way flow of information from germ plasm to soma. Once formed, the organism merely transmits germinal material unchanged to its offspring and then dies.

In the early discussions of this theory (translation 1891–2), Weismann had not linked his concept of a separate germ-line to the cell nucleus, but he soon became aware of van Beneden's work and realized that the behaviour of the chromosomes during reduction division and fertilization matched his prediction of how the germ plasm would be transmitted. From this point on he had, in effect, matched Galton's view of heredity. He had separated the transmission of hereditary characters from the process of growth by claiming that the organism transmits but does not manufacture the units of the germ plasm. He had also decided that the germ plasm consists of discrete units which are combined and recombined through sexual reproduction but which never blend together. Again like Galton, however, he did not appreciate the possibility that unit characters might be traced through a number of generations by simple breeding experiments with discontinuous variants. In part this was because Weismann pictured the germ plasm as a complex system in which characters would interact in various ways. But it may also have been a by-product of his commitment to Darwin's theory of evolution by the natural selection of minute differences between individual organisms.

Ernst Mayr (1985) has hailed Weismann's contributions to nineteenth-century evolutionism as second only to those of Darwin himself. The concept of hard heredity implicit in the theory of the germ plasm clarified the need for natural selection by making the inheritance of acquired characters implausible. Since the soma cannot affect the germ plasm it transmits, any characters it acquires which were *not* produced from the information stored in the germ plasm cannot be reflected in the production of new hereditary units or ids to be transmitted to future generations. Weismann also performed a famous experiment to back up his contention that the inheritance of acquired characters cannot occur. He cut off the tails of generation after generation of mice and showed that there was no tendency for the mutilation to be inherited. In his own theory, the germ plasm for the production of tails was stored in the reproductive system, not – as in pangenesis – produced in the tail itself. The experiment was intended to disprove both Lamarckism and pangenesis by confirming that a change imposed on one part of the body is never reflected in the germinal material transmitted from generation to generation. From this point on, Weismann hailed the 'all sufficiency of natural selection' (1893b). Variations were produced solely by changes arising within the germ plasm without response to the demands of the organism. Selection was the only way in which the concentration of new characters could be directed within the evolving population.

Weismann thus joined Pearson as a staunch defender of the selection theory in the confused decades at the end of the century. His 'neo-Darwinism' – in which all evolutionary mechanisms except natural selection were repudiated – was presented as a logical refinement of the position advocated by Darwin himself. Yet far from encouraging support for Darwinism, Weismann's dogmatism seemed to alienate the vast majority of biologists who still felt that the inheritance of acquired characters could not be dismissed so easily. They pointed out that the experiment on the inheritance of mutilations could not disprove the possibility that usefully acquired characters might eventually become part of the species' inheritance. Since cytologists such as Hertwig distrusted the germ plasm theory, it was easy to present Weismann's whole position as a complex hypothetical edifice built on very shaky foundations. Weismann himself made things worse by postulating an elaborate internal structure for the germ plasm in which the ids (the units on the chromosomes) were composed of smaller determinants responsible for shaping the growth of the embryo. He also put forward a theory of 'germinal selection' in which the determinants struggle amongst themselves to see which will gain most control over the developing organism (translation 1896). In the end he was forced to concede that this theory left open the possibility of nonadaptive and hence non-Darwinian evolutionary trends.

It is also significant that Weismann's last survey of evolution theory (translation 1902) still contained chapters on recapitulation, the regeneration of lost organs, and other topics that had been of central interest to the old developmental world view. For all that the basic concept of the germ plasm helped to set the scene for the emergence of rigidly hereditarian views, Weismann can more plausibly be seen as the last representative of the old tradition rather than a pioneer of the new. He had tried to destroy Lamarckism, pangenesis and the analogy between memory and heredity, but he had not realized that if these concepts were abandoned the logic of both recapitulationism and the integration of growth and inheritance into a single field of study was threatened. The ideas of discrete hereditary units and of a one-way flow of information from the germ plasm to the growing organism would eventually be incorporated into classical genetics. But Weismann himself did not forsee the value of breeding experiments with discontinuous variations and paid little attention to the rediscovery of Mendel's work. He had used the latest developments in cytology to clarify the concept of hard heredity that lay implicit within Darwin's theory of natural selection, but he could not relate to the breeding experiments that would eventually open up a new science of heredity.

An important influence on the final version of Weismann's germ plasm theory was the work of the Dutch botanist Hugo De Vries. In his 1889 book *Intracellular Pangenesis* (translation 1910a) De Vries set out to modify Darwin's theory in a way that would make it consistent with the new cytology (Robinson, 1979, ch. 8). He argued that the really useful aspect of pangenesis was the concept of particles responsible for producing each character in the new organism. Darwin's belief that such particles circulate around the body was obviously unacceptable, but the essence of pangenesis could be preserved by supposing that the hereditary units existed within the cell. In fact, De Vries had abandoned what to Darwin had been the central feature of the theory, the assumption that the gemmules are budded off from the various parts of the body. De Vries' 'pangenes' were discrete units, each representing a single hereditary character, somehow encoded within the physical structure of living protoplasm. The total number of pangenes might be relatively small, the complexity of the organism being a result of their interaction in the growth process.

De Vries went on to insist that the pangenes were located in the cell nucleus, and that they could only multiply through cell division. He accepted that the units were grouped along the chromosomes, but believed that they passed from the nucleus into the surrounding cytoplasm of the cell in order to become active in shaping growth. The chromosomes thus did not play a major role in De Vries' thinking, perhaps because as a botanist he found Weismann's absolute distinction between germ plasm and soma to be unworkable. The fact that most plants can produce buds capable of being grown as separate organisms showed that the germinal material is not confined solely to the reproductive system. Where Weismann postulated a gradual dividing up of the inherited germ plasm among the cells of the developing organism, De Vries insisted that each cell must retain the full complement of pangenes, even if in a latent form.

In the long run, the most significant point of divergence between De Vries and Weismann lay in their choice of future research programmes. As we have already seen, Weismann's strongly Darwinian views on evolutionism, coupled with his belief that the germ plasm is composed of vast numbers of determinants, ensured that he was not especially interested in the problem of discontinuous variations. De Vries' claim that the number of pangenes was relatively small allowed him to accept that at least some characters might exist in distinct forms with no possibility of blending. Later on he would advance his widely popular 'mutation theory', in which new characters – perhaps even new subspecies – were produced by the spontaneous appearance of new hereditary units. But for the time being he was aware that many plant species exhibit discontinuous variation, i.e. alternative forms of the same characteristic which are not linked by a range of intermediates. He realized that by cross-breeding the different forms one might gain some insight into the way in which the pangenes responsible for the alternative characteristics are transmitted from one generation to the next. In the course of the 1890s he undertook a series of hybridization experiments with maize and other species, from which he eventually obtained what we now recognize as Mendelian ratios. De Vries thus became one of the 'rediscoverers' of Mendel's laws, and we must now turn to see how this new line of research transformed the debate over the nature of heredity.

5

Mendel's Contribution

Most histories of genetics agree that the new science was established through the 'rediscovery' of Gregor Mendel's work in 1900 (e.g. Dunn, 1965; Sturtevant, 1965; Carlson, 1966). As a result, Mendel has been awarded the status of a 'hero of discovery'. The basic generalizations upon which genetics is founded are still known as Mendel's laws, and the story of his life has become part of the mythology of science, endlessly depicted in popular books and even in TV programmes. A whole industry of historical research has grown up around the study of Mendel's life and work, with its own journal, *Folia Mendeliana*, issued by the Mendelanium at the Moravian Museum in his home town of Brno. And yet Mendel is an anomalous hero of science because his discovery was ignored during his own lifetime. His posthumous fame has been enhanced by the aura of tragedy generated by the image of a brilliant scientist who could not get his contemporaries to understand the significance of what he was doing. The initial lack of recognition creates problems, however, for scientist-historians seeking to evaluate Mendel's impact, since they have to explain why so clear an exposition of the laws of inheritance failed to impress at the time. In the last decade or so, a new and more sophisticated approach to the problem of Mendel's role in the development of modern biology has begun to emerge, exposing the myths which have built up around his name and trying to clarify the exact nature of his contribution.

The central puzzle in the traditional account arises from the apparent incompatibility between the clarity of Mendel's insight and the failure of his contemporaries to appreciate what is so obvious to the modern reader. If a great discovery really does involve stripping the veil of obscurity from a corner of nature's activity, surely everyone ought to be able to see the truth once it has been pointed out to them. Two explanations for the blindness of

Mendel's fellow biologists are interwoven in the classical story of his discovery. Perhaps his relative obscurity prevented his paper from being read widely enough to have any real impact. Or perhaps he was 'ahead of his time' because the existing body of thought on the question of heredity was so confused that no one could believe that a single experiment held the key to a whole new theoretical insight. Only when everyone else's ideas had become clearer was it possible to see that such simple breeding experiments could solve apparently complex problems. This clarification had occurred by the end of the century, by which time the value of a Mendelian approach had become so obvious that three biologists independently rediscovered the laws of heredity and hailed the now-dead Mendel as the true pioneer.

Analysis of the 'long neglect' of Mendel's paper is complicated by the fact that in the tradition established by his biographer, Hugo Iltis (translation 1932) and continued in Loren Eiseley's history of evolutionism (1958), Mendel's laws could have eliminated the critical weakness in Darwin's formulation of the theory of natural selection. If only Darwin or one of his followers had found Mendel's account, Jenkin's 'swamping' argument could have been undermined by simply pointing out the particulate nature of heredity. This interpretation relies on the 'too obscure' theme to explain why the Darwinians did not, in fact, make the obvious connection, but it assumes that they would have been able to understand Mendel's ideas if they had access to his papers. The alternative 'time was not ripe' explanation assumes that anyone reading the paper would have been unlikely to appreciate its true significance. Our study so far has endorsed the latter style of explanation, but suggests that something more than a simple clarification of ideas was needed before Mendel's work could be appreciated. The prevailing view of inheritance was 'confused' because it rested on a theoretical structure which conceptualized the problem along lines quite alien to post-Mendelian thought. Only by destroying this earlier paradigm would the significance of Mendel's discovery become apparent.

On this model, neither the Darwinians nor their opponents would have understood Mendel because their thinking did not make crucial distinctions which lie at the heart of the modern approach to heredity. Even the next generation of biologists, including Weismann, still tended to think of reproduction and growth as parts of the same biological process. The notion of hard heredity paved the way for a recognition of distinct hereditary units, but in the end Mendel's laws would become the focus for a new science because some early twentieth-century biologists agreed to stop worrying about growth and concentrate on the transmission of characters from parent to offspring. The decision to focus on this episode as the critical step in the foundation of genetics is confirmed by the work of several historians who have begun to question whether Mendel himself was engaged in quite the 'forward-looking' research programme attributed to him by later biologists. If Mendel was conducting hybridization experiments for a purpose in which the identification of hereditary units did not figure as the central goal, then perhaps it is less surprising that his contemporaries did not view him in the same light as the 'rediscoverers' of a later generation. Mendel's role in the history of genetics must be reassessed completely if it turns out that he would not have been counted as a 'Mendelian' by his posthumous followers, had they realized what his real intentions were.

This reassessment of Mendel's position depends on the recognition that hybridization experiments were traditionally used to investigate not the laws of heredity but the problem of the origin of species. It has long been recognized that hybridization had been studied intensively in the period before Mendel began his work, but it is now especially important for us to understand the purpose of these earlier experiments. Instead of looking for 'precursors' of Mendel's discovery, we must ask if the existing tradition might have created an entirely non-Mendelian (in the modern sense of the term) context within which even Mendel himself might have been led to investigate the phenomenon.

Early Hybridization Studies

Since Mendel's laws were discovered by cross-breeding varieties of plants with significantly different characteristics, historians of genetics have naturally looked to earlier hybridization studies in the hope of identifying naturalists who had stumbled across the phenomena that Mendel himself would clarify. There was substantial interest in plant hybridization from the mid-eighteenth century onwards, and it is not difficult to find examples of breeders who noticed the occasional instance of dominance or segregation of characters. In general, though, historians have had to admit that these earlier studies came nowhere near anticipating Mendel's theoretical analysis. The classic account of plant hybridization before Mendel by Roberts (1929) notes that two separate traditions existed side by side: the horticulturalists, and the more 'scientific' students of the phenomenon. The horticulturalists were practical men who wanted to know how new and commercially useful varieties could be created and fixed 'in the breed'. The more academic students of hybrids were inspired by a particular problem which had surfaced in the early eighteenth century: whether or not new species could be produced by cross-breeding existing ones.

Although this latter group of studies were perhaps more carefully organized and reported, they addressed a theoretical issue which can now be seen as significantly at odds with the search for laws of heredity. Crossing closely related species may certainly reveal facts about how characters are inherited, but if the central concern is to see if hybrid species can be formed, it is unlikely that the laws of inheritance will feature strongly in the results that get reported. The horticulturalists were interested in how new characters could be fixed within the species. Their work was thus more likely to bear directly on the question of heredity, but lack of scientific training would have limited the role they could play. In modern terms, it was the horticulturalists not the species-hybridizers who were more directly concerned with heredity. But if Mendel's own work on heredity was a by-product of a deeper interest in the origin of new species, then the species-hybridizers established a major component of the intellectual context within which Mendel conceived his experiments. Instead of stressing only the theoretical innovations that Mendel introduced, we shall have to accept that his work – however well-planned – may have been intended as a contribution to a debate which has nothing to do with the emergence of Mendelism as a theory of heredity.

Among horticulturalists looking for new varieties, the crossing of existing strains was viewed as the starting point from which new characters could be isolated and eventually fixed in the breed. Hybridization was thus a subject of considerable practical interest, and a number of investigators conducted careful experiments to see how divergent characters intermingle. Thomas Knight, President of the Horticultural Society of London, undertook a series of experiments with peas in the early nineteenth century, anticipating Mendel's realization that the number of distinct varieties in this plant made it an ideal subject. This work was followed up by Alexander Seaton and John Goss, who also reported isolated examples of the phenomena we now know as dominance and segregation (Stubbe, 1972, ch. 6). At a primitive level, these men had observed the effects that Mendel would study more carefully, but they made no effort to determine numerical ratios that would allow the inheritance of unit characters to be traced over a number of generations.

The results of these practical studies were certainly known to the naturalists of the early nineteenth century, but the topic of hybridization had already acquired a distinct theoretical character thanks to the speculations of one of the previous century's greatest biologists, Carolus Linnaeus. As the founder of a new system of classification, Linnaeus gained an international reputation. To most of his followers he was known as a supporter of the traditional belief that species are permanent elements within the divine plan of creation. But in the course of his long career, he had begun to have second thoughts on this topic and eventually came to believe that many species have been formed naturally in the course of time. Linnaeus' principal hypothesis to explain the origin of species was not transmutation but hybridization (Roberts, 1929; Glass, 1959b). In the 1750s he and his pupils reported a number of newly discovered species which they interpreted as having been formed recently through the hybridization of previously existing species. In his *Disquisitio de Sexu Plantarum* of 1756, Linnaeus went so far as to suggest that in the creation, God had only formed a single member of each plant genus – the range of modern species in each genus had been formed over the intervening centuries by hybridization. He believed that the female parent determined the essential character of the hybrid, the male effecting only superficial changes which were nevertheless permanent in the new species.

Linnaeus' suggestion was not as radical as a complete theory of evolution, since it still required the original form of each genus to be designed by God. But a number of later naturalists responded to his claims by investigating the question of inter-specific hybrids more carefully. In general they were anxious to disprove Linnaeus' assertions by showing that hybridization between species cannot occur in a way that would establish an entirely new form. The first important contribution was made by Joseph Gottlieb Kölreuter, who reported a series of observations in the course of the 1760s (reprinted 1893; Olby, 1985; Mayr, 1986). Kölreuter believed that nature is a harmoniously designed system and opposed Linnaeus because he felt that the production of new species would upset the pattern established by the Creator. He was committed to the fixity of species, but opposed the theory of pre-existing germs (see Chapter 2). He regarded the offspring as a true blend of parental characters produced by a mixing of seminal fluids. To ensure the stability of the natural system, however, it was essential that the characters of two distinct species should not be able to mix through cross-breeding. In normal circumstances, geographical separation ensures that cross-pollination does not occur between closely related plant species. Hybrids *can* be produced by human interference, but Kölreuter was determined to show that the intermediate form is normally sterile and thus cannot serve as the parent of a new specific type.

Kölreuter was clearly puzzled when he discovered that there were occasional exceptions to this rule, and equally puzzled when he found that the second hybrid generation – far from maintaining the blended character of the union – showed a considerable range of variation. On several occasions he observed the discontinuous inheritance of characters, but he was convinced that these must be exceptions to the normal rule of blending. Kölreuter thus reported observations that can be translated into Mendelian terms, but he was unable to think in terms of particulate inheritance because his theory of generation presupposed the smooth mixing of seminal fluids. His real aim throughout his experiments was to disprove Linnaeus' claim that new species could be produced by hybridization, and in the end he concluded that he had demonstrated the impossibility of this ever occurring under natural conditions.

Kölreuter's results were by no means universally accepted, and in the 1820s Carl Friedrich von Gaertner began another extensive series of hybridization experiments. In 1837 Gaertner's account of his work was awarded a prize by the Dutch Academy of Sciences for the best study of how new species and varieties could be produced by artificial fertilization. A revised version of his submission eventually appeared in 1849 under the title *Versuche und Beobachtungen über die Bastarderzeugung im Pflanzenreich*, which both Darwin and Mendel acknowledged as the most extensive source of information then available on the subject. Like Kölreuter, Gaertner carried out many crosses in the tobacco genus, which we now know to have a complex genetic constitution quite unsuitable for demonstrating Mendelian effects. In the course of investigations on other genera, Gaertner occasionally observed dominance and segregation, but offered no systematic explanation of these effects. In general Gaertner – like most of his contemporaries – believed that the whole character of the species participated in the combination of essences brought about during hybridization. He did not accept that individual characters might be traced through a series of generations as a means of throwing light on inheritance. Whatever the origin of Mendel's interest in hybridization, his experiments with peas would offer an entirely new methodology.

Mendel's Life and Work

The orthodox image of Mendel as the prophet of genetics, unheeded by his own contemporaries, originated at the time of the 'rediscovery'. Early Mendelians such as William Bateson stressed the 'modernity' of Mendel's approach to heredity. They assumed that the chief aim of Mendel's research was the elucidation of laws of inheritance applicable to all species. On this interpretation, Mendel was bitterly disappointed when, at Nägeli's suggestion, he tried to repeat his experiments with Hawkweed (*Hieracium*), only to find that in this unsuitable subject the hybrids did not show the segregation of characters already demonstrated in peas. Later geneticists updated their image of Mendel even further – to them it was unthinkable that he could have planned his experiments without the idea of paired hereditary particles, so he must have effectively anticipated the concept of the gene. Mendel's biographer Hugo Iltis (translation 1932) extended this image and reinforced the claim that the 'long neglect' of his paper had been due to his contemporaries' primitive knowledge of the cytological processes responsible for inheritance. Writing at a time when genetics had begun to link up with a revived Darwinism, Iltis identified the myth of Mendel the pioneer geneticist with the claim that the problems of nineteenth-century Darwinism would have been solved if only his work had been recognized. From this time onward, most historians of genetics have assumed that Mendel himself was interested in evolution as well as heredity.

Mendel's career fits neatly into this image of the frustrated genius. Numerous accounts of his life are available: in addition to Iltis (1932) see for instance Olby (1985) and Orel (1984). Johann Mendel was born into a family of peasant farmers in 1822, but received a good education once his abilities had been recognized. Failing health and family problems led to him being admitted as a novice at the Augustinian monastery in the Moravian town of Brünn (now Brno, Czechoslovakia) under the adopted name 'Gregor'. To facilitate a career as a teacher he studied at the University of Vienna for two years, 1851–3, his most important lecturers being the biologist Franz Unger and the physicist Christian Doppler. From Unger he learnt about the latest developments in cytology and must have become aware of his teacher's controversial support for the general idea of evolution. He also learnt about the hybridization experiments of Kölreuter and Gaertner, which were of concern to Unger because they appeared to contradict evolutionism by endorsing the stability of species. From Doppler, Mendel absorbed the mathematical approach to scientific problems, and it is generally assumed that his novel experimental technique represents a pioneering attempt to apply the mathematics of the physicist to a biological problem. A nervous illness prevented him from completing the examinations at Vienna, so he returned to Brno as a supply teacher in the local school.

Over the next few years Mendel performed his classic experiments on hybridization, using peas grown in the monastery garden. The results were described to the Brno Natural History Society and published by the Society's journal in 1866 (translation in Bateson, 1902 and in Stern and Sherwood, 1966). Although Brno was by no means a cultural backwater, the paper appeared to attract little attention. Carl Nägeli was the only major biologist with whom Mendel was able to interact, through a series of letters written between 1866 and 1873 (also in Stern and Sherwood, 1966). Many of these letters discussed new experiments with *Hieracium*, on which Mendel published another paper in 1870. The results bore no resemblance to those obtained with peas, leading to the popular assumption that Nägeli's encouragement of the *Hieracium* work led Mendel to become disillusioned with his whole research project. In 1868 Mendel was elected abbot of the monastery and soon found that administration absorbed all his time. His experiments ended in 1871 and in his final years he became embroiled in a bitter dispute with the government over the monastery's tax bill. He died in 1884, mourned by the local community but ignored by the scientific world.

The traditional view is that Mendel's experiments with peas were undertaken to show that in this species characters can be treated as distinct units or elements which are transmitted unchanged from parent to offspring. Mendel surveyed the available true-breeding varieties of peas and identified seven character differences whose inheritance could be tested by hybridization:

round or angular seeds;
yellow or green seeds;
white or grey-brown seed coats;
smooth or ridged pods;
green or yellow pods;
axillary or terminal flowers;
tall or short (overall height of plant).

Using careful techniques of artificial fertilization, Mendel performed the appropriate crosses and showed that in the first hybrid (F1) generation, one character out of each pair appears in all the hybrid plants. The other character has *apparently* disappeared, a situation which Mendel described by saying that one character in the pair was 'dominant' over the other. In the case of height, when tall and short plants were crossed all the hybrids were tall, so tall is the dominant character. Mendel then self-fertilized the hybrids to produce the second hybrid (F2) generation, where he observed the famous 3 : 1 ratio. What Mendel called the 'recessive' character reappeared in one quarter of the F2 plants, giving a ratio of three dominant (e.g. tall) to one recessive (e.g. short).

Mendel explained this phenomenon of the segregation of characters by suggesting that the first hybrid generation combined the hereditary potential of both characters, the recessive being masked or hidden but nevertheless available for transmission to the next generation. In the F2 generation there is an independent assortment of the two characters so that all four possible combinations occur. If we denote the tall character by T and the short by t, the hybrid will possess the potential for both, Tt, but will appear tall because the T character is dominant. In the second generation we have all four possible combinations:

$$TT \quad Tt \quad tT \quad tt$$

of which the first three will show the dominant character, while the last, tt, reveals the recessive which had been masked in the F1 plants. In this model, the appearance of the plant is *not* a guide to its hereditary constitution: pure TT appears the same as mixed Tt, but only the latter has the power to transmit the recessive character to future generations. Further experiments with two and three character-pairs allowed Mendel to show that the pairs segregated independently of one another.

Even within the conventional interpretation of Mendel's programme, certain puzzling features have been noted. The independent assortment of characters observed in the later experiments occurs (in modern terminology) because the genes responsible for each character-pair lie on different chromosomes. If they did not there would be a 'linkage' between characters controlled by the same chromosome. The pea has seven chromosomes, and it is thus popularly supposed that Mendel chose the maximum number of independently segregating characters available to him. Unfortunately, it is now known that the characters he chose do *not* all lie on separate chromosomes, so Mendel ought to have discovered linkage (Blixt, 1975).

A rather different problem was first pointed out by R. A. Fisher (1936), who noted that in statistical terms Mendel's results are a little too good to be true. He would have had to be very lucky indeed to get ratios as close as he did to the theoretical prediction. Various explanations have been offered to account for Fisher's analysis, including an overenthusiastic assistant and unconscious bias in counting. The most obvious point to note, however, is that the character-pairs are not all as distinct as popular legend supposes (Root-Bernstein, 1983; Piegorsch, 1986). Some are what may be called 'fuzzy sets', which leave room for a certain amount of observational bias if they are artificially broken up into supposedly distinct categories. True-breeding discontinuous characters are not, in fact, very common in nature, which is why Mendel's results could so easily be dismissed as anomalous. Even the pea, while offering better experimental opportunities than most species, is by no means perfect. It would take a great leap of the imagination to believe that the pea experiments provided a model which – allowing for various complicating factors – would explain the whole phenomenon of heredity.

A far more serious problem for the historian is posed by a series of recent studies which, taken together, present an entirely new picture of Mendel's intentions. Without challenging the innovative character of his experiment on peas, a number of historians have begun to argue that the orthodox interpretation of his research programme has been influenced by the geneticists' desire to see Mendel as a pioneer student of their discipline. The traditional image of Mendel is a myth created by the early geneticists to reinforce the belief that the laws of inheritance are obvious to anyone who looks closely enough at the problem. Suspicions were first voiced in the 1960s, but it was the almost simultaneous publication of two papers by Augustine Brannigan (1979) and Robert Olby (1979, reprinted in Olby, 1985) that made a revaluation inescapable. Olby's article bore the provocative title 'Mendel no Mendelian?', his intention being to suggest that a careful reading of the pea-hybridization paper reveals no sign of certain key concepts of twentieth-century Mendelism that Mendel is popularly supposed to have anticipated. The 'rediscoverers' had read a good deal of their own thoughts into Mendel's paper, and later geneticists have been content to accept the claim that he intended to convey all the ideas that his posthumous disciples attributed to him. Olby also noted that the real purpose of Mendel's experiments was not to create a new model of heredity, but to resolve the old question of whether or not new species can be produced by hybridization.

Most geneticists have assumed that Mendel could not have analysed his results in the way he did without the concept of paired material particles responsible for the control of a particular character. We now know that chromosomes exist in homologous pairs, so that a normal cell contains two genes for each character, an egg or sperm cell only one. It is assumed that Mendel must have anticipated this fact and gone on to recognize that each hereditary particle or gene can exist in one of two forms (now known as alleles) corresponding to the character-pairs studied in his experiments. In such a theory, the gene controlling the height of the pea plants must exist in two forms, tall (T) and short (t), the former being dominant. The pure-bred varieties from which Mendel started would have the genetic constitution TT (appearing tall) and tt (appearing short); each have both genes of the pair identical. Since each parent contributes only one of its genes to the offspring, the F1 plants must have the genetic constitution Tt (also appearing tall). The F2 plants will have the four possible combinations listed above, of which one quarter (tt) will appear short. The 3 : 1 ratio thus seems to follow naturally from the idea that the segregation of characters is controlled by hereditary particles existing in pairs.

This is how modern geneticists interpret Mendel's experiments, and they have naturally assumed that Mendel himself must have analysed his results in the same way. But Olby points out that the description of the pea-hybridization experiments contains no explicit reference to the concept of paired material particles equivalent to the genes of modern Mendelism. Although Mendel discussed the cytological implications of his results, he nowhere indicates that the germ cells contain one character-particle and the fertilized egg two. H. Kalmus (1983) suggests that Mendel's ability to think in terms of paired *characters*, but not paired *particles*, reflects his early training in scholastic or Aristotelian philosophy. Far from being a materialist committed to the notion of hereditary particles coded for particular characters, Mendel adopted the scholastic view that character is determined by opposing essences. His concept of dominance may have come from the equally Aristotelian distinction between the actual and the potential. On this interpretation, Mendel's experiments arose from the fruitful interaction of two methodologies: the mathematical technique of the modern physicist was combined with a much older philosophical tradition to give a new approach to the analysis of biological character.

Equally significant is Olby's reinterpretation of Mendel's primary intentions. Geneticists have traditionally assumed that the pea-hybridization experiments were performed in the hope of un-covering universally valid laws of inheritance – hence Mendel's 'disappointment' when the crossing of different *Hieracium* species produced no sign of segregation. Olby suggests, however, that Mendel's interest in hybridization stemmed from his awareness of the existing debate over the possibility that new species might be produced by crossing existing forms. He set out to show that characters can be transmitted independently to the hybrid because he thought that new combinations of characters – corresponding to new varieties and perhaps new species – might thus be generated. The crossing of pea varieties would be relevant to the production of new species because Mendel saw no reason to draw an absolute distinction between varieties and species. When he found that crosses between *Hieracium* species showed no segregation, only the production of constant hybrid forms (now explained as the result of apomixis), he saw this as even better support for the view that new species could originate in this way.

An even stronger statement of this thesis comes from L. A. Callender (1988), who insists that Mendel's intentions have been misinterpreted because of the widespread assumption that he favoured the general theory of evolution. Callender sees Mendel as an opponent of evolutionism who was trying to defend Linnaeus' alternative theory of the production of new species by hybridization against the attacks of Kölreuter and Gaertner. Mendel advanced no universal laws of inheritance because he believed that hybridization occurs in two quite different ways. The peas illustrated the phenomenon of 'variable' hybrids, while *Hieracium* showed the production of 'constant' hybrids that were potentially new species. The restricted 'laws' of inheritance shown in peas were thus of less interest to Mendel than his *Hieracium* work. Far from being dragged unwillingly to *Hieracium* as an experimental subject by Nägeli, Mendel set out deliberately to study what he regarded as the more interesting kind of hybridization shown by this genus. His intention was to show Nägeli that new species in the genus were produced by hybridization not by transmutation, and his disappointment lay in the fact that he was unable to convince the more influential biologist of this before he was forced to give up his work due to other pressures on his time.

The fact that Mendel had a general theory of the production of new species by hybridization is confirmed by his references to the struggle for existence – a struggle he conceived not in Darwinian terms, between individuals, but as a process taking place between the new species formed by crossing. In his last letter to Nägeli he pointed out that changing conditions ultimately render the male organs of a plant sterile, so that hybrids are formed through fertilization with pollen from other species. He continues:

If this were actually the case, the naturally-occurring hybridizations in *Hieracium* should be ascribed to temporary disturbances which, if they were repeated often or became permanent, would finally result in the disappearances of the species involved, while one or another of the more happily organized progeny, better adapted to the prevailing telluric or cosmic conditions, might take up the struggle for existence successfully and continue it for a long stretch of time, until finally the same fate overtook it. (Stern and Sherwood, 1966, p. 102)

On Callender's reading, the 'better adapted progeny' are new species produced by hybridization of the old ones before they die out. The study of segregation in peas, however careful and potentially fruitful, was thus merely a side issue in Mendel's campaign to show that Linnaeus' theory of hybrid species offered an alternative to the new evolutionism.

The 'Long Neglect'

The new interpretation of Mendel's theory and of his intentions has obvious implications when we turn to consider why his work on peas failed to attract the attention of his contemporaries. On the traditional assumption that his main purpose was to use hybridization as a tool to investigate the inheritance of distinct characters, his analysis (even if it does not use the concept of paired particles) was aimed at exactly the problem for which the rediscoverers of 1900 would develop similar techniques. Since the laws of inheritance do, in fact, work discontinuously – and since existing theories of heredity were obviously unsatisfactory – the modern geneticist is inclined to assume that Mendel's work *could* have been accepted at the time as the key to a theoretical breakthrough. If it was not, then we need to ask why his contemporaries failed to recognize his paper's potentially revolutionary implications. Obviously, if Mendel's intention was *not* to advance a new theory of inheritance, this whole issue will have to be viewed in a very different light.

The orthodox history of genetics accepts the intrinsic plausibility of Mendel's 'laws' and offers two explanations for his contemporaries' indifference. Mendel was ignored either because his paper was too obscure to attract attention or because he was 'ahead of his time' in suggesting a simple answer to what everyone else saw as a complex problem. The obscurity of the journal in which the paper was published – the *Verhandlungen des naturforschenden Vereines in Brünn* has often been noted. Like the transactions of many local scientific societies it was routinely circulated to other societies and libraries and was theoretically available to Darwin or anyone else who might have been able to make use of the paper. Olby and Gautry (1968) have identified eleven references to the paper in the literature before 1900, so it was not entirely ignored. Yet these references indicate only routine citations within the literature on hybridization; there is no evidence that anyone saw Mendel's work as a breakthrough in heredity. MacRoberts (1985) argues that scientists do not, in any case, pick up ideas from searching the published literature. They almost always rely on personal contacts within the profession for news of what is going on. Mendel did not belong to such an informal network, and his one real contact with access to the world of professional science, Nägeli, did not think his work of sufficient importance to draw other biologists' attention to it.

There is thus a sense in which Mendel's paper was neglected rather than ignored: with the exception of Nägeli, most biologists would simply have been unaware of its existence. Even so, one could argue that a single contact with the world of professional science should have been enough to gain the paper a hearing, if its message was really comprehensible at the time. Our perception of the situation changes dramatically, however, once we accept that Mendel's chief interest was the production of new species by hybridization. Perhaps Mendel himself made no serious effort to promote his 'laws of heredity' as the basis for a new science, because he did not see those laws as being universal and because they represented only a by-product of his main research programme. If Mendel was engaged in trying to convince Nägeli that new species are generated by hybridization rather than evolution, there is no reason why Nägeli should have been expected to see the experiments with peas as a breakthrough in an entirely different area. Mendel regarded his research as incomplete and thought that further experiments on *Hieracium* (which he had no time to perform) would be needed. Later geneticists have misconstrued the situation because they assume that Mendel himself was trying to promote a theory similar to their own. If Mendel did not attach such a degree of importance to his pea hybridization experiments, we should hardly be surprized that others did not realize their potential. The theory of hybrid species was a very minor challenge to the growing authority of Darwinism. Mendel's contribution to this declining research programme was hardly likely to attract much attention, especially as his results with peas were not as interesting (in this context) as the *Hierarcium* hybrids.

Obviously the biologists who 'rediscovered' Mendel's laws in 1900 saw his experiments as a valuable contribution to the problem they had just begun to take up. They had good reason to apply the results with peas to the question of heredity rather than of hybridization. The concept of new hybrid species was now almost totally forgotten, while heredity had emerged as a key biological problem. Mendel's experiments seemed to offer clear support for ideas towards which the rediscoverers themselves were now groping their way. For his paper to become useful, however, it had to be read in a context very different to that in which it had been written. We can now see that there was no chance whatsoever of anyone in the period immediately following the paper's publication being in a position to anticipate this alternative context. To extract the kind of insight that the rediscoverers found in the paper, the biologists of the 1860s would have had to realize that the behaviour of Mendel's 'variable' hybrids held the key to the whole problem of heredity. No one was in a position to do this because there was no existing framework into which a model of discontinuous character-inheritance could have been fitted. Even if it had been better known, Mendel's paper would have been ignored because the interpretation we read into it today would have been unthinkable to the biologists of the time.

We have already seen that the prevailing 'developmental' view of generation made it virtually impossible for anyone in the middle decades of the nineteenth century to appreciate that a study of pure heredity had any value (Sandler and Sandler, 1985). Mendel could ignore the question of growth because – unlike Darwin and the other evolutionists – he had an entirely different view of the origin of species. New forms are produced by mixing old characters, not by a process of variation that generates new ones. But without a study of growth, Mendel's analysis of heredity (if it was applied in that area) was incomplete by the standards adopted by almost all his contemporaries. If Mendel was not trying to promote a new theory of heredity, then he cannot have been 'ahead of his time'. To the extent that his work could be used as the basis for a new model of heredity, it addressed the problem in a way that was out of touch with the prevailing conceptual framework. Only when that framework changed would it become possible for a new generation of biologists to look back and see that the inheritance of characters in peas offered a model upon which to build a whole new theory of heredity.

6

The Emergence of Mendelism

Since Mendel's work was without immediate impact, the emergence of a distinct science of Mendelism (soon to be called genetics) dates from the rediscovery of his laws of inheritance by Carl Correns, Hugo De Vries and Erich von Tschermak in 1900. Soon after their announcements, Mendel's discovery was taken up by William Bateson as the basis for a new science of heredity. By the end of the first decade of the new century, Mendelism was already well on the way to becoming a distinct area of biology – although important theoretical developments still had to be made before the 'classical genetics' of the 1920s and 30s could be established.

We have already seen why Mendel's approach had been unlikely to attract much attention in the mid-nineteenth century, and have noted that Mendel himself may not have intended to promote his work on peas as the key to the whole problem of heredity. Obviously the situation had changed by 1900 so that the idea of unit characters could gain a foothold – but which of the factors discussed in previous chapters played the most effective role in paving the way for the rediscovery of Mendel's laws? The ideas of Galton and Weismann had certainly created a framework within which some biologists had begun to think in terms of 'hard' heredity, and the cytologists had begun to promote the view that particles in the cell nucleus might be responsible for controlling the production of individual characters. Some figures associated with the rediscovery – De Vries is the best example – had participated in the cytological debates and may thus have been primed to think of characters being fixed by units in the germ plasm. But the claim that characters were somehow 'preformed' in the nucleus was rejected by most embryologists, who still felt that inheritance was a subsidiary issue within the general problem of development. Several early Mendelians, including Bateson, came from an embryological background and consistently repudiated any link between unit characters and hypothetical particles in the nucleus.

Some other factor must thus be invoked to explain the sudden enthusiasm for breeding experiments designed to reveal the transmission of unit characters. One possibility is the resurgence of interest in the idea of discontinuous or saltative evolution. In reaction to the neo-Darwinism of Weismann and Pearson, many biologists began to express opposition to the whole idea of gradual evolution. Even Galton thought that new species were formed by the sudden appearance of entirely new characters. A number of figures associated with the rise of Mendelism, including De Vries, Bateson and, later on, T. H. Morgan, began from a similar position. Because they thought evolution worked through the sudden creation of new characters by forces within the organism, they were inclined to assume that such characters would be inherited as units once they were created. It was eventually realized that a discontinuous model of heredity does not imply that evolution must also be discontinuous, since a particular character may be influenced by a number of Mendelian factors. But the early Mendelians did not explore this possibility and some of them remained hostile to Darwinism for several decades. Mendel's technique of using breeding experiments to trace the transmission of unit characters was thus hailed as the key to a new science of heredity at least in part because of the mistaken belief that the unit-character concept was essential for a theory of saltative evolution. The link to the cytologists' theory of germinal units within the chromosomes would only be forged by later geneticists.

Did the rediscovery of Mendel's laws spark off a dramatic revolution in scientific thinking on heredity? It certainly introduced a new factor into the debates, and one that would eventually play a vital role in eliminating the old ideas of soft and blending inheritance. But if there was a revolution it took well over a decade to be completed, so the early version of Mendelism can be seen at best as only a first step in the right direction. In fact, we shall see that many biologists continued to doubt the new theory's value long after the geneticists themselves thought the revolution was over. Historians now recognize that the development of genetics was far more complex than a simple recognition of the particulate nature of heredity. The new theory took some time to emerge, and many of the early Mendelians did not foresee concepts we now take for granted. Their Mendelism was only a half-way house on the road to modern genetics, leaving important conceptual developments to be made before the revolution was completed. The immature state of early Mendelism should encourage us to take a more sympathetic view of those biologists who did not recognize the potential of the new approach.

The emergence of Mendelism was more than a conceptual revolution: it required the establishment of an entirely new branch of biology devoted to the study of topics that had hitherto been parcelled out among other fields. The creation of genetics was thus an event within the development of the scientific profession. We need to ask how the early practitioners of the new discipline established a niche for themselves – and to what extent their success threatened the position of rivals seeking a share of limited professional opportunities. The politics of a scientific revolution can be very complicated, and the circumstances surrounding the 'Mendelian revolution' are no exception. We are now beginning to realize that the new science succeeded at least in part because it was able to tap a new supply of resources by appealing to agricultural interests which could be expected to benefit from improved breeding programmes.

The early phase of the revolution can be divided into two episodes: the actual rediscovery and the subsequent promotion of Mendel's laws as the basis for a new science of heredity. The two events are distinct if only because two of the rediscoverers – De Vries and Tschermak – played no role in the creation of the new discipline. Indeed, the traditional story of a simultaneous redis-covery of Mendel's laws by Correns, De Vries and Tschermak is now surrounded by controversy. Doubt has been thrown on the claim that De Vries and Tschermak independently recognized the significance of what later became known as Mendel's laws. Para-doxically, these doubts raise the prospect that Mendel's long-neglected paper played a vital role in clarifying the issues – even if Mendel himself had not intended his work to be interpreted in this way. The rediscoverers may have had complex motives for acknowledging Mendel's priority, but his analysis of how unit characters behave was to become an important factor in the campaign to use the experimental evidence as the basis for creating a new science.

The Rediscovery

The classic account of the simultaneous rediscovery of Mendel's laws is given by Roberts (1929). Here we are presented with a clear impression that each of the three men independently discovered evidence for, and recognized the significance of, the segregation of characters. Each of them saw their work as the foundation for generally valid laws of heredity, and only afterwards came to realize that the now-dead Mendel had anticipated their experiments and their conclusions over thirty years previously. The events are routinely described as a classic example of independent discovery, illustrating how obvious the laws of nature become once an area has been cleared of conceptual ambiguities. The implication of the story in its original form is that once biologists began to look for evidence that characters could be inherited as units, they were bound to discover the laws governing how those units are transmitted.

Historians have now thrown considerable doubt upon the viability of this story. The atmosphere had certainly changed since Mendel's time, allowing biologists to become receptive to the basic idea of discontinuous heredity. But recognition of the laws now attributed to Mendel was by no means as immediate as the notion of triple rediscovery implies. The claims of one rediscoverer, Tschermak, have now been almost universally rejected, while controversy surrounds the extent to which De Vries was able to solve the problem before he read Mendel's paper. The traditional story may thus be an artefact of the geneticists' imaginations: they have exaggerated the coincidences between the three biologists' work in order to enhance the view that Mendel's laws are obvious to anyone who looks at the problem carefully. Each of the three rediscoverers obviously had his own reasons for stressing his independence of Mendel's earlier work. By claiming to have become aware of Mendel's paper only *after* making the discovery for themselves, they attempted to retain some degree of creativity in the eyes of posterity. If Mendel's paper (whatever its author's intentions) actually helped the rediscoverers to clarify their own thinking on heredity, the new concepts may not have been as obvious as the later geneticists would like us to suppose. It has also been suggested that Correns deliberately invoked Mendel's name once he realized that De Vries had already published on segregation. If this is so, then the image of Mendel as the unlucky precursor may have begun as an attempt to head off a priority dispute among the early students of discontinuous heredity.

Erich von Tschermak's reference to Mendel in his paper of 1900 was long accepted as sufficient to establish him as one of the rediscoverers. This claim was repudiated by Curt Stern, however, who insisted that Tschermak had not understood Mendel's laws and refused to include Tschermak's paper in his Mendel sourcebook (Stern and Sherwood, 1966). Olby's account of the rediscovery (1985) points to Tschermak as an example of a biologist who interpreted what we now regard as Mendelian phenomena within a pre-Mendelian concept of heredity. Simply recognizing Mendelian ratios was not enough to guarantee an appreciation of the underlying mechanism by which geneticists now explain these results.

Of the three rediscoverers, the botanist Carl Correns is the least controversial. It is widely accepted that he came across evidence for the 3 : 1 ratio in segregation while investigating maize hybrids in the late 1890s. Correns himself claimed that the latest developments in cytology had encouraged him to think of paired characters and thus led him towards a new interpretation of his experimental results. Certainly the hereditarianism lying at the heart of the theories proposed by Galton and Weismann must have focused attention on characters that might remain unchanged over many generations. Yet Olby (1985) points out that Weismann refused to accept Mendelism, suggesting that by itself cytology would have encouraged the idea of multiple, not paired, hereditary units. Whatever the changes in the general background during the decades leading up to 1900, Mendel's concept of paired characters still had something to offer. The question is whether or not that concept was independently reinvented by Correns and De Vries.

Olby suggests that even Correns may not have appreciated the full significance of his results until after he had read Mendel's paper. Certainly, he proceeded only slowly to prepare a paper of his own for publication, until he read De Vries' first report and realized that he no longer had priority. Mendel's analysis of how character-pairs behave over several generations was superior to that developed by any of the rediscoverers, and may thus have played a vital role in shaping the ideas of the first geneticists. Correns himself continued to doubt that the laws were universally valid, as Mendelians such as Bateson soon began to claim. However paradoxical it may seem, the analysis of inheritance in peas that Mendel neglected in favour of what he regarded as the more interesting phenomenon of constant hybrids was to play a significant role in bringing the idea of paired characters to the attention of early twentieth-century biologists.

The Dutch botanist Hugo De Vries is by far the most controversial figure in the rediscovery episode. Like Correns, he too maintained that he had observed segregation before reading Mendel's paper, and the possibility that the discovery could have emerged directly from De Vries' own research has been defended by Lindley Darden (1976). But Correns was suspicious of De Vries, whose later recollections of the event were by no means consistent. In recent years Kottler (1979), Campbell (1980), Meijer (1985) and Olby (1985) have all maintained that De Vries did not independently discover the phenomenon of segregation because the interpretation offered in his papers of 1900 depended heavily on his reading of Mendel's account. The first paper, published in French, makes no reference to Mendel, although it uses his terminology (De Vries, 1900a). The second paper (1900b) is a longer account in German which does refer to Mendel, although some critics claim that the references could have been added in proof after De Vries saw Correns' response to his first paper. Meijer (1985) argues that De Vries probably saw Mendel's paper some years earlier but did not understand it; only in 1900 did he reread the paper and see that it offered a new way of analysing the results he had been obtaining for some years.

Although De Vries may not have appreciated the possibilities of the character-pair concept until his second reading of Mendel, his own research had led him to a situation in which he was able to think of characters as distinct hereditary units. His *Intracellular Pangenesis* of 1889 (see Chapter 4) supported the view that characters were determined by material particles or 'pangenes' in the cell nucleus. His research during the 1890s was concerned with the nature of variation, of which he distinguished two kinds: fluctuating variability due to changing numbers of pangenes for the same character and 'species-forming' variability where an entirely new kind of pangene/character appeared. This was consistent with Galton's view that continuous variation was supplemented by saltations in the formation of species. De Vries was now convinced that evolution should be studied by showing how new characters are introduced into a population. To demonstrate that new characters are formed as units he tried to show that a character from one species can be transferred to another by hybridization. This led him to the phenomenon of segregation which he eventually interpreted in the terms proposed by Mendel.

It is significant, though, that De Vries only began to use Mendelian terminology in 1900 and abandoned it soon afterwards. For him, at least, the rediscovery of Mendel's laws was a flash in the pan. He soon began to argue that the laws were of little significance because they threw no light on the origin of new characters. This was to become the central theme of De Vries' later work, within which he constructed his 'mutation theory' of character formation (translation 1910b; Allen, 1969; Bowler, 1978, 1983). Although the concept of mutation was subsequently incorporated into genetics, De Vries' original 'mutations' were conceived as large saltations that established a new species at once. His evidence from the evening primrose, *Oenothera lamarckiana*, was eventually shown to be due to a process unrelated to what we now know as genetic mutation. The popularity of De Vries' theory certainly helped the cause of Mendelism, since many biologists felt that discontinuous evolution required a discontinuous model of heredity. But De Vries himself did not believe that his mutations obeyed Mendel's laws, and it was left for others to ensure that the insights promoted in his and Correns' papers of 1900 would be applied in the field of heredity.

The Early Mendelians

The most active proponent of Mendel's laws as the basis for a new theory of heredity was William Bateson. Within a few years, Bateson had emerged as the head of a small but active group of researchers at Cambridge, although he was forced to defend the theory against opposition from the biometrical school. Far from being welcomed with open arms, Mendelism sparked off bitter controversies – often over issues which seem futile today because we can see that both sides had valuable insights to offer. The differing perspectives were exaggerated by personality conflicts, and by philosophical and ideological antagonisms. Historians have also begun to recognize the professional constraints within which the biologists had to operate. Kevles (1980) points out that both Bateson and his chief rival, Karl Pearson, were professionally insecure at the time. The antagonisms were less marked in America, where the scientific community was expanding and thus offered more room for new arrivals. Even here, though, biologists such as T. H. Morgan resisted the introduction of Mendelism for some years. In both countries, geneticists succeeded in creating a place for themselves at least in part by appealing to agricultural interests, presenting Mendelism as the key to improved breeding programmes.

Bateson began his career as an evolutionary morphologist studying the origin of the vertebrates. He soon rejected this quest as fruitless and began to argue that evolution must be investigated by trying to uncover the process by which new characters are produced through variation. He became convinced that characters are produced discontinuously, by saltations. His *Materials for the Study of Variation* (1894) launched a bitter attack on the Darwinian theory, both for its emphasis on continuity and for its reliance on adaptation as the sole directing agent of evolution. Bateson suggested that new characters are produced by biological processes and persist whether or not they are of any use to the organism. As part of his research programme he began to hybridize related varieties in the hope of understanding how the unit characters distinguishing them (which he believed to have been produced by saltations) would behave. He was clearly prepared to think of characters as discrete units, although he did not independently recognize Mendelian ratios. According to the biography prepared by his wife, Bateson read Mendel's paper on the train to London and immediately changed the lecture he was to give at the Royal Horticultural Society to include a reference to Mendel's laws (B. Bateson, 1928, p. 73). Robert Olby (1987a) has now shown that it was probably De Vries' first paper that Bateson read on this occasion, so he did not know of Mendel's work at the time. His conversion to Mendelism was thus a somewhat more gradual process over the next two years.

By 1902, Bateson's *Mendel's Principles of Heredity* included both a translation of Mendel's paper and a defence of the claim that his laws would turn out to be universally valid (Carlson, 1966, ch. 1 and 2; Cock, 1973; Darden, 1977). In the same year he introduced the term 'allelomorph' (subsequently abbreviated to 'allele') to denote the alternative characters in segregating pairs. He coined the term 'genetics' in 1905 and launched it at an international congress the following year. With a small group of followers including R. C. Punnett, he began a series of experiments to extend the range of phenomena to which Mendelism could be applied. Nevertheless, Bateson's situation at Cambridge remained precarious: he had no professorial chair and had to depend on research grants from the Evolution Committee of the Royal Society. Eventually he moved to the John Innes Horticultural Institute at Oxford, while Punnett obtained the first chair in genetics at Cambridge in 1912.

Jan Sapp (1987) argues that Bateson's efforts to establish a new science of genetics involved a deliberate challenge to the authority enjoyed by traditional areas of biology. By presenting genetics as an experimental science, biologists from other backgrounds were to be excluded from the field of heredity. Naturally enough, non-experimental biologists resisted this challenge and tried to limit the new science's territory. The dispute with biometry is an obvious product of this professional rivalry, but other areas of biology such as palaeontology refused to acknowledge the authority of genetics and resisted its influence for several decades. Olby (1987b) points out that genetics benefited from a new attitude in Edwardian Britain, where for the first time it became accepted that the government should actively support useful scientific research. Since Bateson could offer the hope that genetics would help to improve breeding stocks for agriculture, he was able to tap new sources of funds and thus escape the restrictions that would otherwise have been imposed by professional rivalries.

Although Bateson deliberately set out to create a new science of heredity based on Mendelism, he never lost sight of his original interest in discontinuous evolution. His *Problems of Genetics* of 1913 continued the attack on Darwinism, but shows that Bateson had by now become suspicious of the kind of saltations proposed in De Vries' mutation theory. While convinced that evolution was not a continuous process, he felt that biologists were still a long way from understanding how it worked. In an address delivered the following year, he even suggested that evolution never produces genuinely new genetic characters: what we take to be 'new' characters appear only because masking genes are at last destroyed by degenerative mutation (Bateson, 1914; Bowler, 1983). His followers, however, had fewer reservations, and Punnett's popular little text *Mendelism* (2nd edn, 1907) openly supported 'mutations' as the source of the new characters which provide saltative evolution.

Bateson's new science thus allowed him to continue participating in an old debate. Although De Vries abandoned Mendelism and Bateson abandoned the mutation theory, the majority of biologists saw the new theory of discontinuous heredity as support for the increasingly popular idea of discontinuous evolution. Mendelism thus became caught up in the debate over Darwinism. Karl Pearson and W. F. R. Weldon had used biometry to support Darwinism and had rejected the saltative evolutionism proposed in Bateson's *Materials*. Inevitably they saw Mendelism as a new move in the campaign to discredit their position on evolution and responded accordingly (Froggatt and Nevin, 1971; Provine, 1971; Norton, 1973). Weldon wrote actively against Mendelism, claiming that Mendel's own experimental results were suspicious. Pearson and Weldon still defended Galton's law of ancestral heredity and denied the Mendelians' claim that only the parental generation need be taken into account in calculating the offspring's inheritance. They pointed to the many characters that show a continuous range of variation rather than the distinct forms studied by Mendel. Bateson was at first prepared to concede that continuous variation might be a separate phenomenon not governed by Mendel's laws, but under the pressure of Weldon's criticism he soon moved to argue that all evolutionarily significant characters are controlled by the discontinuous mode of heredity. The debate reached heights of bitterness which held up the spread of Mendelism for several years. The situation eased somewhat when Weldon died in 1906. By 1911 both Bateson and Pearson had gained a more secure professional situation and adopted a somewhat less combative stance.

The intensity of the debate was certainly generated by personality clashes, but there were deep divisions between the two sides at many levels. The biometricians were committed to the study of heredity and variation as a mathematical problem, while Bateson's mathematical abilities were limited and he preferred to concentrate on simple experiments. Pearson adopted a positivist philosophy which made him suspicious of any attempt to postulate an underlying mechanism for heredity, while Bateson preferred a simpler, inductivist methodology (Norton, 1975; Cock, 1973). Bateson also adopted what has been called a 'conservative' philosophical position (Coleman, 1970), and this has been linked to his distrust of biometry's tendency to see everything in terms of whole populations (Mackenzie, 1982). As we shall see (Chapter 8) Bateson was suspicious of the social implications that Pearson drew from his interpretation of heredity. The philosophical and ideological distinctions between the two sides have led Mackenzie and Barnes (1979) to see the debate as an example of how different, but equally valid, scientific objectives can be established within a single field of study. Against this, Roll-Hansen (1983) argues that there *was* an objective criterion for deciding which side was right and that the Mendelians eventually triumphed by experimental demonstrations of their position's superiority.

Modern geneticists will probably welcome the assertion that the early Mendelians were justified in rejecting the biometricians' objections, but it is clear that the Mendelians' view of heredity and variation was deficient by today's standards. As Mayr (1973, 1982) points out, both sides in the debate were confused over the relationship between evolution and heredity, since neither realized that evolution can be continuous even though heredity works by a discontinuous or Mendelian process. As early as 1902, G. Udney Yule showed that Mendelism was consistent with the continuous range of variation that the biometricians saw as the basis for gradual evolution. If several Mendelian factors influence the same character, their effects will blend to give an apparently continuous range. Bateson himself realized that more than one factor might be involved (Olby, 1987a), but his commitment to saltative evolution made it impossible for him to admit that species might evolve continuously. Only in the 1920s did it become possible to create a science of population genetics by using the biometricians' statistical techniques to apply the laws of genetics to the more complex case of heredity in whole populations.

Although Bateson recognized the need to treat characters as discrete units (or at least combinations of discrete units) he refused to adopt what later geneticists would regard as the most obvious explanation: that each character pair is controlled by paired particles in the cell nucleus, the alternative forms of the character being somehow encoded in the material structure of the particles. Coleman's (1970) interpretation of Bateson's conservative philosophy reveals him as an opponent of materialism who preferred a holistic view in which waves or similar physical functions spread

throughout the whole cell were responsible for transmitting hereditary information. For Bateson, it was pure simple-mindedness to
assume that each character was linked to a material particle or
'gene'. Darden (1977) shows that Bateson also refused to acknowledge the growing split between genetics and embryology.
Although in practice he had given up embryology for the study of
how characters are transmitted, he never abandoned the hope that a
complete theory of heredity would still explain how the characters
are actually produced in the growing organism. Almost to the end of
his career, Bateson resisted what was to become the dominant
feature of classical genetics.

Most historians of genetics accept that Bateson and his followers
founded the first effective school of Mendelism and thus paved the
way for later developments. Attention then switches to America,
where T. H. Morgan and his school took the next logical step by
linking Mendel's laws to the chromosome theory of inheritance to
create 'classical' genetics (see next chapter). The fact that this neat
sequence is to some extent an artefact of the historians' imagination
is obvious once we recognize that Morgan himself refused to accept
Mendelism for some years, while Bateson resisted the introduction
of chromosome theory into Britain. The artificiality of the traditional story becomes even more apparent when we turn to look at
other European countries. Each nation's scientific community
reacted in a different way both to Mendelism and to classical
genetics. Intellectually and professionally, biologists responded to
the new initiatives in ways that seemed most appropriate in their
own environment. Neither Mendelism nor the chromosome theory
were guaranteed success, and the failure of some countries to accept
the new science requires us to define rather more carefully what it
was that ensured success elsewhere. Even with hindsight, we cannot
pick out an 'obvious' line of theoretical development, since
countries which contributed little at one stage were able to use their
alternative approaches to the study of heredity to good effect later
on.

The hostility of the biometricians had placed Bateson in a
position where he had to work up Mendelism as the foundation of a
new science or give up all hope of gaining a position within the
scientific establishment. His situation at Cambridge, coupled with
his success in tapping new sources of funding, enabled him to found
an effective school of Mendelism despite the criticism arising from

other areas of biology. Immediately across the English Channel, however, the fate of Mendelism was quite different.

The new genetics gained scarcely any foothold in the French scientific community, a fact often linked to the continued preference of French biologists for Lamarckian views on evolution (Buican, 1984). But the situation was in fact more complex (Burian, Gayon and Zallen, 1988). French biologists evaluated Mendelism from within existing conceptual frameworks established by nineteenth-century physiologists and microbiologists such as Claude Bernard and Louis Pasteur. Information about the work of Bateson and other early Mendelians was widely disseminated in France, and a number of biologists did experiments to test the validity of Mendel's laws. But hardly any took up Mendelism as the basis for active research programmes, nor was there any significant effort to establish a role for the new science in the university curriculum. The one biologist who did important Mendelian work, Lucien Cuénot, discouraged his students from getting involved. The possibility of creating a new research tradition was seen as unrealistic since genetics appeared to turn its back on important issues that French biologists refused to abandon. Even Cuénot's real interest lay in physiological genetics – he wanted to know how the Mendelian factors act to control the growth of the organism so that a particular character is produced. Many biologists dismissed Mendelism as relevant only to trivial characters, and it was widely accepted that the cytoplasm as well as the cell nucleus was responsible for controlling development. It was from this background, not from classical genetics, that a later generation of French biologists would become leading figures in the emergence of molecular biology.

In Germany, Mendelism fared much better but still did not pave the way for acceptance of the chromosome theory (Harwood, 1984). A new field of *Vererbungswissenschaft* (heredity-science) emerged, with journals of its own and exponents in several universities and in the Kaiser Wilhelm Institut für Biologie. Geneticists such as Carl Correns and Richard Goldschmidt gained international reputations. And yet the German science of heredity was both professionally and intellectually more diffuse than the genetics of the English-speaking world. The German biologists, like the French, refused to give up the hope that a complete science of heredity would also explain how the characters are produced in the developing organism. They too continued to insist that the cyto-plasm played a vital role in heredity and thus resisted the emphasis placed on the chromosomes by classical geneticists of the American school. Mendelian effects were studied, but were always seen as but one part of a much wider research programme.

The Danish botanist Wilhelm Johannsen introduced concepts that have been incorporated as vital components of modern genetics. He pioneered the distinction between the 'phenotype' and the 'genotype' and actually coined the term 'gene'. The phenotype is the organism's physical character, while the genotype is its genetic constitution. Organisms with the same genotype may differ to some extent in their phenotype, because environmental influences during growth may affect the way in which the genetic characters are expressed. But this somatic variation is not inherited: only the genotype is relevant for determining the characters that can be transmitted to the next generation. This is a vital distinction, since it eliminates any possibility of a Lamarckian effect. Environmental factors can influence the growth of the individual, but not the transmission of genetic factors. Johannsen used this distinction to great effect in establishing genetics as a distinct area of biology. The phenotype was proclaimed irrelevant for the study of inheritance. Only the genotype mattered, and the geneticist was the only person who could tell how the genes were transmitted (Sapp, 1987). Yet as Churchill (1974) and Roll-Hansen (1978) have shown, the meaning of the terms 'genotype' and 'phenotype' has changed since their first introduction, since Johannsen originally defined them in terms of populations, not individuals. The modern use of the terms evolved as geneticists sought to establish a definition of heredity that would allow their new science to become the undisputed authority in a particular area of biology.

Johannsen had been deeply influenced by Galton's view of heredity, but he came to believe that the biometricians were wrong to depict a population as representing a continuous range of hereditary variation. Experiments with beans allowed him to isolate a number of 'pure lines' existing within the natural population, each pure line showing no inheritable variability and thus being resistant to further selection. The phenotype of the population varied continuously for a character such as size, but the genotype varied discontinuously since it consisted of a number of pure lines. Mendelism was no longer restricted to the study of distinct characters: even continuous variation could be seen as the product of a limited number of genetic factors, their effects blended together by trivial somatic variation. Selection could pick out the distinct pure lines, but could have no effect on the purely somatic or phenotypic variation among individuals with the same genotype.

Because Johannsen's beans were self-fertilizing, he had little interest in the original Mendelian phenomenon of segregation. His work did much to establish the concept of the gene as an absolutely stable entity, transmitted from one generation to the next entirely free from environmental influence. But he was also interested in the possibility that entirely new genetic factors might occasionally be produced by what is now known as mutation. Apparently new pure lines had emerged spontaneously in the course of his experiments, suggesting that the genetic variability of a population could be increased discontinuously. His work was thus taken as support for De Vries' view that new varieties are produced instantaneously by saltations. The relatively trivial differences between the pure lines did, however, force the Mendelians and saltationists to scale down the size of the discontinuities by which evolution was supposed to take place (Van Balen, 1986).

Johannsen's concept of the pure line as a fixed genetic type seems to reflect the old typological view that species are based on clearly defined organic forms. Although he introduced the term 'gene' and believed that the corresponding entity had an objective existence, he shared Bateson's reluctance to admit that it might correspond to a material particle in the chromosome coded to produce a particular character. For Johannsen, the gene was forever unobservable and probably consisted of stable energy-states within the organism as a whole. The views expressed by Bateson and Johannsen thus represent an intermediate stage between nineteenth-century biology and classical genetics: both linked Mendelism to saltative evolution and neither was willing to visualize transmission in terms of material particles. Yet by the standards of French and German biologists, their position was far more central to the development of ideas on heredity. Instead of dismissing their form of genetics as incomplete, we need to recognize that the 'next step' in the advance of genetics – the link with chromosome theory – was a unique event within American biology, an event which required the deliberate repudiation of ideas central to the thinking of many Europeans and the temporary neglect of issues that would regain significance with the emergence of molecular biology in the 1940s.

A small number of American biologists began to work on Mendelism soon after 1900 and the theory rapidly gained in popularity. Here there was no debate with biometry: some biologists did statistical work on variation, but they were not tied down by the uncompromising support for Darwinism expressed by Pearson and Weldon. In fact, biometry, Mendelism and the mutation theory all benefited from their ability to satisfy the increasing interest of agriculturalists in scientific breeding (Kimmelman, 1983; Paul and Kimmelman, 1988). As Bateson found in Britain, here was a new source of funding available to anyone who could claim to provide an experimental approach to traditional problems. The American Breeders' Association supported work on Mendelism and the mutation theory after 1903, with the popularity of genetics increasing as biologists began to realize that there was something wrong with De Vries' theory. American geneticists also benefited from the fact that the academic community was expanding rapidly at the time, making it easier for biologists to establish new courses and even new departments in areas that seemed to offer practical results.

Yet the development of Mendelism in America was not without problems. At Harvard, William E. Castle was inspired by Bateson's work to demonstrate that albinism in mice behaved as a Mendelian recessive factor. But Castle soon decided that the unit characters of Mendelism were not absolutely distinct: he believed that they could 'contaminate' one another in a way that obscured the effects of segregation. Over the next few years he defended this view against all comers, including his Harvard colleague E. M. East (Carlson, 1966, chs 3 and 4). East argued that *characters* cannot always be regarded as units – it is the Mendelian factors that are unitary and several of these may affect a single character. One gene may be able to modify the way in which another is expressed, thereby helping to produce a continuous range of variation in a large population. But East too resisted the idea that the gene might correspond to a single material entity. In his view the concept was merely a mathematical abstraction used in analysing the results of breeding experiments.

Even in America there were many biologists who actively

opposed the early version of Mendelism. As we have already seen (Chapter 4) embryologists such as E. G. Conklin were opposed to the 'preformationism' of the germ plasm concept and believed that the inheritance of characters was only a subsidiary element within the general problem of explaining how the organism grows to maturity. A similar view was held at first by the biologist who subsequently went on to establish the link between Mendelism and chromosomes, Thomas Hunt Morgan (Allen, 1978, 1986). Morgan was an embryologist committed to studying the forces that shape the organism's growth. Early work on the regeneration of lost organs convinced him that this characteristic could not have been built up gradually by evolution, and he attacked the Darwinian theory vigorously in his *Evolution and Adaptation* (1903; Allen, 1968; Bowler, 1978, 1983). Morgan was convinced that new characters appeared by saltations or De Vriesian mutations and became established whether or not they were beneficial to the organism. In the period before 1910 Morgan also criticized both Mendelism and the chromosome theory of inheritance on the grounds that they seemed to imply preformationism, thereby ignoring the process of development by which characters are actually produced in the growing organism. Such theories focused attention on the structure of the hereditary material and evaded the main problem of development by simply postulating the existence of particles corresponding to the adult character. Morgan's conversion to both Mendelism and the chromosome theory will be the subject of the next chapter.

The emergence of Mendelism had clearly introduced a new factor into the debate over heredity, strengthening the hand of those biologists who believed that inheritance cannot be influenced by environmental forces. As yet, however, the new theory was highly controversial and was unable to precipitate a dramatic conversion of the whole scientific community to hereditarianism. The inspiration for many early Mendelians was the concept of discontinuous evolution, and in this respect the criticisms of the biometricians were not unjustified. Nor had the theory been able to link up with the most obviously hereditarian model for the physical processes involved in reproduction. With a few exceptions such as De Vries, many early Mendelians were hostile to the concept of preformed hereditary particles. The ambiguity of the unit-character approach made it difficult to relate the simple Mendelian experiments to the more complex phenomena of variation seen in large populations. It also blunted the search for a mechanism of transmission that would allow this process to be differentiated from growth. Johannsen's work helped to reinforce the general hereditarian claim that characters are transmitted independently from one generation to the next, but until his 'genotype–phenotype' distinction took on its modern form, embryologists such as Morgan would continue to regard transmission as a field unworthy of study in its own right. The division between embryology and genetics had begun to emerge in practice, but had not yet been defined in principle. Around 1910 the distinctions we take for granted today at last began to be clarified, especially through the realization that the behaviour of the chromosomes during reproduction parallels the effects described by Mendel's laws. Classical genetics could then begin to emerge as the dominant form of hereditarian theory – at least in those countries where the chromosome theory achieved complete dominance.

7

Classical Genetics

By the standards that have become common among English-speaking geneticists, the Mendelism of early figures such as Bateson was at best only a half-way house toward the establishment of true, i.e. 'classical', genetics. The refusal of Bateson and many early Mendelians to accept that the gene might be a material unit in the chromosome robbed their theory of much of its potential value. They had no *mechanism* of heredity and were thus forced to rely on simple breeding experiments to reveal the behaviour of Mendelian factors. Often they failed to distinguish between the character in the adult organism and the Mendelian factor which produced it. Modern genetics only emerged when T. H. Morgan and his colleagues finally identified the Mendelian factor as a material unit or 'gene' in the chromosome, thereby opening up a cornucopia of insights that would explain a host of complex phenomena and clear up once and for all the confusion over unit characters and genetic factors. Whatever the complexities of the growth process by which characters were expressed in the adult organism, genetics was now in a position to assert that all the characters lay somehow predetermined or encoded in the structure of the chromosomes.

The parallelism between the segregation of Mendelian characters and the behaviour of the chromosomes during meiosis and fertilization was noticed by a few biologists very soon after 1900. It is easy to argue with hindsight that the parallel was so obvious that a link between Mendelism and the chromosome theory of inheritance was inevitable. Those geneticists who objected to the chromosome theory were put under increasing pressure to explain why their philosophical and methodological arguments should stand in the way of exploring the possibility of synthesis (Van Balen, 1987). The problem with this interpretation is that it assumes the emergence of an autonomous science of genetics to be an inevitable consequence of the need to achieve clarity in the conceptual scheme by which the phenomena of inheritance are analysed. We have to believe that embryologists such as Morgan were forced by the very logic of their situation to abandon their objections to preformationism and accept that the problem of growth (the expression of the genes' potential) could be set aside to let geneticists get on with the job of studying heredity (the transmission of genes from one generation to the next).

There can be little doubt that the establishment of classical genetics did involve a suspension of interest in growth, and hence the creation of a rigid distinction within what had hitherto been perceived as a unified field of study devoted to problems of 'development'. But it is precisely the establishment of an autonomous science of heredity that requires us to view the whole process as something more than a clarification of conceptual issues. Historians now realize that the creation of classical genetics (like the original creation of Mendelism) represented a bid for scientific authority by a group of biologists who wished to define a particular area of conceptual 'territory' as their own (Allen, 1986a; Sapp, 1987). Even in the rapidly growing American scientific community there was much to be gained in the shape of research funds and academic positions by presenting a new theoretical approach as the key to the solution of practical problems associated with animal and plant breeding.

The possibility that what we now take to be a clarification of the issues was originally inspired by motives of professional advancement may strike many geneticists as an unwholesome challenge to their science's legitimacy. Yet the very fact that the chromosome theory was not universally accepted in the 1920s and 30s – especially by French and German biologists – should alert us to the fact that the geneticists' refusal to take an interest in how genes express themselves in growth was a repudiation of issues that many biologists quite legitimately felt to be of major importance. The research opportunities opened up by so drastically reducing the width of focus in the area of inheritance were enormous, but important issues had to be put on one side to allow this concentration of effort, and many biologists were not prepared to go along with so one-sided a programme. The classical geneticists were able to stifle dissent within the American scientific community, but elsewhere the topics they repudiated were still studied and were

often used as vehicles for the presentation of ideas at variance with
the rigid hereditarianism of the chromosome theory. In particular,
the claim that the cytoplasm plays a vital role in growth (and hence
in inheritance) was used to promote a more flexible view of the role
played by heredity in determining an organism's character.

One by-product of classical genetics was a softening of the
anti-Darwinian prejudices expressed by most early Mendelians. It
was now realized that mutations merely fed new characters (often
only slight modifications of existing characters) into the population
– they did not establish new species instantaneously as De Vries had
claimed. Even Morgan at last became willing to admit that adaptive
value would determine whether or not a mutated gene would spread
into the population. But to cope with the genetics of large
populations, biologists needed techniques quite different to those
appropriate for laboratory breeding experiments. The biometricians had, of course, pioneered the use of statistics for analysing the
variability of large populations, but their commitment to Darwinism had led them to reject Bateson's Mendelian programme.
Eventually younger biometricians such as R. A. Fisher began to
tackle the job of applying statistical techniques to the genetics of
large populations. In so doing they healed the rift between
Mendelism and Darwinism and paved the way for a revitalized
theory of natural selection to become the basis of the modern
'synthetic' approach to evolution. Significantly, however, those
biologists who objected to the absolute dominance of the chromosome theory of heredity also tended to favour the retention of
non-Darwinian mechanisms of evolution. The classical geneticists
dismissed these anti-Darwinians as the last vestige of an outdated
developmental tradition – but the opponents of Mendelism and
Darwinism saw themselves as preserving a role for developmental
factors that had been ignored in the rush to oversimplify the
problem of heredity.

The Chromosome Theory

In 1911 Wilhelm Johannsen visited the United States, and his paper
'The Genotype Conception of Heredity' was published in the
American Naturalist. Jan Sapp (1987) regards this paper as a crucial
move in the drive to establish genetics as an autonomous science,

the sole repository of knowledge about heredity. Here Johannsen used his concept of the genotype to dismiss traditional studies of the phenotype (bodily characters) as irrelevant for the understanding of heredity. Since the phenotype could be influenced by all sorts of non-heritable environmental factors, only the geneticist could see through to the underlying content of the germ plasm which alone was responsible for determining the character of future generations. At one stroke, all rival disciplines such as biometry and embryology were to be stripped of their authority in the field of heredity. Garland Allen (1986a) accepts that Johannsen's visit probably exerted a decisive influence on the one biologist who was to be most effective in establishing the new discipline in America: Thomas Hunt Morgan (for biographies of Morgan see Allen, 1978; Shrine and Wrobel, 1976).

Yet Johannsen repudiated the concept that Morgan and his colleagues were to use as the intellectual tool for securing the dominance of genetics. Like Bateson, he consistently refused to accept the possibility that the gene might be a material structure in the chromosome coded to produce a particular character. Several biologists had, however, already suggested that Mendelian segregation might result from the separation of the paired chromosomes during gamete formation, followed by the joining of chromosomes from egg and sperm at fertilization (Baxter and Farley, 1979). If the alternative forms of the alleles were somehow preformed or encoded in certain parts of the chromosomes, the fact that each parent contributed one of its chromosomes to the offspring's pair would explain the transmission of characters according to Mendel's laws. The separation and recombination of the chromosomes in the reproductive process showed an exact parallel with the transmission of Mendelian characters. Cytology would thus offer an explanation of the Mendelian effects, provided one accepted that the chromosomes are indeed the bearers of heredity. This point was noted by the American biologist Walter S. Sutton (1902) and endorsed by Theodor Boveri (1904), whose earlier work had helped to establish the continuity of the chromosomes through the reproductive process (see Chapter 4).

In America, Edmund B. Wilson of Columbia University began to test what he was later to call the 'Sutton–Boveri' hypothesis. He was now increasingly convinced that chromosomes transmitted from the parents play a vital role in predetermining the characters of the growing organism. In particular he began to argue that, in insects at least, sex is determined by an accessory chromosome which had been discovered in the sperm-producing cells of the males. Crucial work supporting this view was performed by Nettie M. Stevens in 1905 (Brush, 1978). From a position of initial scepticism, Wilson now became a leading exponent of the view that chromosomes are responsible for transmitting hereditary characters. Wilson recognized that he had thereby endorsed the view originally sketched in by Weismann's definition of the germ plasm as the sole bearer of heredity.

Also at Columbia, T. H. Morgan was aware of Wilson's growing support for the chromosome theory (Roll-Hansen, 1978). At first he remained sceptical, since his original interests in experimental embryology led him to prefer an epigenetic view of development. Both Mendelism and the chromosome theory were rejected as attempts to revive a simple-minded preformationism. Like many embryologists, Morgan confused what Roll-Hansen calls the morphological and the chemical versions of preformationism. He saw the chromosome theory as implying that the adult characters somehow exist in a miniaturized form within the germ plasm. As yet he had not recognized the possibility that the gene might be seen as a material structure capable of influencing growth in a certain way. Under the influence of Wilson and Johannsen, however, Morgan began to adopt a new position in the period around 1910. His original enthusiasm for De Vries' mutation theory led him to study the fruit fly *Drosophila melanogaster* in the hope of extending the evidence for saltative evolution to the animal kingdom. *Drosophila* was to become a favourite organism of the geneticists because it has only four chromosomes which are easily visible. Its mutations played a vital role in converting Morgan to the chromosome theory of heredity.

Morgan never renounced his interest in embryology, but by 1910 he was becoming frustrated by the inability of the experimental approach to yield important results in this area. At the same time he was pragmatic enough to realize that both government and private sources of funding were available for areas of science that could present themselves as sources of useful information (Allen, 1986a). Morgan's switch to research on the transmission of characters certainly allowed him to create a new science that would appeal to agricultural and breeding interests and yet be a respectable academic discipline. To do this he needed a theoretical innovation that would both sustain a major research programme and allow that programme to be depicted as the only legitimate source of new information in its field.

The first steps towards such an initiative came when Morgan became aware that a mutated 'white eye' character had appeared in his *Drosophila* specimens. The discovery helped to disprove De Vries' claim that mutations established new species, since the mutated character could be bred with normal flies and its transmission studied by normal Mendelian techniques. From this point on Morgan began to lose interest in saltative evolution and to recognize the power of Mendelism to reveal how characters are transmitted. The white-eyed character was linked to the sex of the fly, and Morgan began to speculate that its production was somehow determined by the sex chromosome. Shortly afterwards two more sex-linked mutations were discovered, but further breeding revealed an anomaly. If all three characters were controlled by the one chromosome, they should have shown complete linkage, always segregating together. In fact the linkage was not complete; the factors controlling the mutated characters somehow recombined in a way that seemed inconsistent with the chromosome theory.

This incomplete linkage had been observed before by Bateson and Punnett, who had tried to explain it with their non-chromosomal theory of the gene. Morgan chose instead to explore the details of chromosomal behaviour in the hope of finding something that would account for the anomaly. He had already encountered the concept of the 'chiasmatype' introduced by the Belgian cytologist F. A. Janssens to explain the observed intertwining of the chromosomes in the early stages of meiosis. Janssens suggested that the chromosomes broke and reformed at the crossing point, thereby interchanging corresponding segments. Morgan saw that this crossing-over of chromosome segments would explain the incomplete linkage of the sex-linked mutations. Cytology and Mendelism thus came together to provide a fruitful new way of studying the more complex cases of inheritance that were now coming to light.

To exploit the new initiative, Morgan set up the 'fly room' at Columbia University, where vast numbers of *Drosophila* were bred and studied. He was joined by a number of younger biologists, of whom the three most active were Alfred H. Sturtevant, Calvin B. Bridges and Hermann J. Muller. In the period 1910–15 they uncovered a host of chromosomal effects in *Drosophila* and published a flood of reports culminating in their book, *The Mechanism of Mendelian Inheritance* (Morgan, Sturtevant, Muller and Bridges, 1915). Morgan wrote vigorously to popularize their work both among scientists and the general public. By 1915 the Morgan school was riding on the crest of a growing wave of enthusiasm. The Carnegie Institution now stepped in to fund the research, and students were moving out from Columbia to found genetics laboratories throughout the United States and in a number of other countries. In 1933 Morgan was awarded the Nobel Prize for Physiology and Medicine.

The range of techniques and concepts exploited by the Morgan school was enormous (for details see Carlson, 1966; Dunn, 1965; Sturtevant, 1965; Allen, 1978). By studying the frequency of recombination, Sturtevant was able to produce genetic 'maps' indicating the relative positions of the genes on the chromosomes. He also clarified the concept of mutation, showing that a single gene could mutate in various ways to produce a number of different alleles. Mendel's pairs of alternative states had given a simplified picture of the real state of affairs: in fact each gene could exist in a larger number of alternative forms. This point also undermined Bateson's view that the recessive state was merely the absence of the dominant character, with mutations being responsible merely for destroying existing genes. Genetics could now explain the existence of a much wider range of characters in the population.

Morgan and his followers also recognized that a single character can be controlled by a number of different genes, while a single gene may have an effect on more than one character. They also discovered that the position of a gene on the chromosome was important, since its effect could sometimes be modified by neighbouring genes (the 'position effect'). The exact nature of mutation remained a puzzle for some time, however. At last Muller developed techniques for inducing mutations artificially by exposing organisms to X radiation. Under normal circumstances the gene is stable except for a slight chance of mutating to another form, which is often deleterious or even fatal. Muller showed that radiation induced a much higher level of mutation and saw this as a chance to explore the nature of the gene itself. Since the environment normally has no effect on the gene, the often harmful consequences of mutation suggested that the molecular structure which encodes the gene's information can somehow be changed or damaged only by particularly severe external forces. As Carlson (1966) points out, the 1930s saw not a slackening of effort in classical genetics, but a switch of interest to the problem of radiation-induced mutations. Muller was eventually awarded the Nobel Prize for his work in this area (Carlson, 1981).

Sturtevant (1965) presents an idyllic picture of the co-operative atmosphere prevailing in the fly room, but it is clear that the group was not without interpersonal friction. Muller in particular felt himself to be an outsider, and eventually moved on to make an independent career including a period in the Soviet Union. The tension reflected different opinions on scientific, philosophical and political matters. On the scientific front, historians are divided over the creative input of the participants. The orthodox view is that Morgan was the leader both institutionally and intellectually, superintending the creative interaction of his colleagues. Allen (1978) endorses this view to some extent, while noting that Morgan sometimes held back from the bolder speculations of the younger researchers. Carlson (1974, 1981) and Roll-Hansen (1978) present a significantly different interpretation, treating Morgan as a cautious thinker whose role was essentially critical rather than creative. They argue that Morgan often lagged behind the others in accepting the clarifications offered by the new chromosome theory. On this model, Muller emerges as a vital stimulus to the group's thinking. His consistent materialism was a major source of theoretical innovation, although his tactless behaviour made his colleagues reluctant to accept and later to acknowledge his contributions.

Allen (1978) notes that Morgan welcomed the chromosome theory precisely because it provided a material basis for what would otherwise be the purely speculative concepts of Mendelism. Under the influence of his friend Jacques Loeb, he favoured a materialist approach to biology, although he never accepted the thesis that biology can be reduced completely to physics and chemistry (see Loeb, 1912). The classical concept of the gene was thus a triumph of mechanistic biology. Roll-Hansen (1978), on the other hand, treats Morgan as an empiricist who was never really comfortable with Muller's emphasis on the gene as a material determinant. It is certainly true that Morgan took a couple of years to accept the full implications of the gene concept and was at first reluctant even to use the term 'gene'. In later years he returned to his old interest in embryology and became less concerned about the material reality of the gene. According to Carlson and Roll-Hansen, Muller was the real source of the materialist concept of the gene, and the most active in exploiting that concept as a guide to research. Morgan's role was to ensure that the resulting hypotheses were testable in the laboratory.

Putting aside the question of priority, the Morgan group was the source of a new model of heredity in which the emphasis was firmly on the chromosomes as the only vehicle by which characters can be transmitted from parent to offspring. The term 'classical genetics' illustrates the extent to which the new discipline managed to establish itself as the sole source of authority in the field of heredity. Morgan had at first been prepared to allow a role for the cytoplasm in affecting the transmission of some characters, but he soon became an active opponent of cytoplasmic inheritance. This attitude was to become characteristic of classical genetics, responsible for the marginalization of all phenomena that might threaten the claim that the material gene was the sole bearer of heredity. The gene was to be interpreted as a segment of chromosome whose structure was such that it influenced growth in a particular way and was thus able to produce a particular character in the adult organism. The material structure of the gene was a self-replicating unit, transmitting itself and its potentiality unchanged from cell to cell and from parent to offspring. Changes induced in the adult organism by external factors could not affect the gene, so that Weismann's theory of the independence of the germ plasm was effectively vindicated. Mutations were the sole source of new genes and hence of new characters, and these were produced by spontaneous changes in existing genes which could not be influenced by the needs of the growing organism. The inheritance of acquired characters thus became a theoretical impossibility.

On the question of *how* the gene influences the developing organism, the classical geneticists were evasive. From their perspective, it didn't really matter how the effect was produced. If the transmission of characters could be studied by Mendelian techniques, the geneticist was entitled to treat the gene as a fixed unit and to offer his knowledge as a complete answer to the problem of heredity. Defining heredity in the narrow sense of transmission was a crucial part of the geneticists campaign to stake out a territory within which their science would be recognized as the sole source of authority (Sapp, 1987). In principle, of course, they admitted that there was another aspect to genetics that might eventually hope to uncover the process by which the gene did its work in the growing organism. As an embryologist originally, Morgan retained an interest in growth, and his colleagues recognized that the developmental process must intervene between the material structure of the gene and its final expression in an adult character. In practice, however, classical genetics tended to discourage investigation of how the gene works, thus producing a complete split between genetics and embryology – a divorce from which genetics benefited by seeming to represent the more active and successful area of research.

Some critics have argued that for all its ostensible materialism and reductionism, classical genetics used the concept of the gene merely as a convenient symbol for analysing Mendelian and chromosomal phenomena (Stent, 1970). Although linked to the chromosome, the 'material' gene was almost as abstract a concept as that advocated by earlier Mendelians such as Bateson. On this interpretation, the subsequent emergence of molecular biology as the study of what genes are and how they function constitutes a major breakthrough transcending the research tradition of classical genetics. Other historians see the classical geneticists' reluctance to tackle the question of gene expression as merely a tactical manoeuvre designed to set on one side questions that could not be answered with the techniques then available. Muller's commitment to the gene as a material particle certainly included the belief that the geneticist should strive to understand the nature of that particle. His work on the effects of radiation was an attempt to study the gene indirectly by investigating the changes it could undergo. He also attempted to define the structural discontinuities along the chromosome, thereby establishing the boundaries of the unit genes, and this was seen as a prelude to an attack on the chemical nature of the units (Falk, 1986). In practice, the geneticists put the question of development to one side, but in principle the work of Muller, at least, can be seen as the retention of an interest in the nature and expression of the gene that would serve as a prelude to the flowering of molecular biology in the 1940s.

Population Genetics

The classical geneticists' concept of 'mutation' differed considerably from that introduced under the same name by De Vries. It was
now clear that the apparently new varieties seen by De Vries in the
evening primrose were due not to genetic mutation but to the
species' hybrid constitution. The true mutations studied in *Drosophila* were new genetic characters introduced into the existing
population, since the affected individuals could interbreed with
normal organisms. Genetics thus no longer supported the view that
new varieties and species could appear by saltations, as Bateson and
Morgan had at first believed. At the same time it was shown that
characters exhibiting a continuous range of variation within a
population are under the influence of several Mendelian factors. As
early as 1902, G. Udney Yule had pointed out that Mendel's laws
were not necessarily incompatible with the phenomenon of continuous variation studied by the biometrical school. If a number of
independently segregating Mendelian factors could influence the
same character, their effects would blend together in the population
to give an apparently continuous range of variation at the phenotypic level. Johannsen's pure lines in beans pointed to a similar
interpretation. By 1910 this interpretation of continuous variation
had begun to gain wider acceptance, thanks to the work of
H. Nilsson-Ehle in Sweden and Edward East in America.

Morgan's co-workers were convinced that Mendelism could
account for continuous variation, and saw mutations as new factors
which merely extended the species' existing range of variability.
They also began to suspect that natural selection might not, after
all, be incompatible with a genetical model of heredity. The larger
mutations observed in the laboratory were almost invariably
deleterious in their effect, so that individuals expressing the new
characters could not be expected to survive in the wild. But some
mutations had a very small effect on an existing character, and
might thus slightly increase or decrease the organism's ability to
cope with the environment. If mutation occasionally produced a
character that was favourable in a certain environment, organisms
expressing that character would do better in the struggle for
existence and would breed more effectively, thereby helping to
establish the mutated character in the population. Evolution would
be a relatively slow process involving the gradual adding together of

a series of small mutations, each so limited in its effect that the overall change would appear continuous. Morgan himself was at first unwilling to admit that he had been wrong to attack natural selection, but by 1916 he had begun to argue more positively for Darwinism.

In effect, Mendelism undermined Fleeming Jenkin's claim that natural selection was ineffective because new characters will be swamped by interbreeding with unchanged individuals. If heredity is particulate rather than blending, a favourable new character will be preserved intact so that its frequency can be gradually increased within the population. Yet this was no simple reconciliation between Mendelism and Darwinism, since genetics had effectively undermined the whole developmental view of heredity. In particular, it had shown that earlier evolutionists (including Darwin) had been wrong to visualize heredity and variation as two antagonistic forces. Instead, heredity and variation *within populations* had to be seen as merely two different aspects of the same process. Where there was significant genetic variability, because a number of genes or alleles affected the same character, variation was only maintained in the population because heredity preserved the variant genes intact from one generation to the next. Mutation explained the origin of the variant alleles and could be expected to supply new characters to extend the range of variability in the future.

The potential for a renewed interest in the theory of natural selection thus existed – yet the model of evolution presented in Morgan's *Critique of the Theory of Evolution* (1916) was still based on an oversimplified form of selection (Allen, 1968, 1978; Bowler, 1978, 1983). Morgan supposed that a harmful gene appearing through mutation would be eliminated immediately by selection, while a favourable one would spread rapidly through the population. There was no room here for a wide range of variation, or the large-scale elimination of unfit individuals. Hermann J. Muller went on to defend a more sophisticated version of this approach (1949), arguing that the vast majority of genes in a population express the 'normal' or 'wild type' character. Only a handful of individuals will carry abnormal genes produced recently by mutation. Selection will keep the frequency of abnormal genes very low, until mutation at last comes up with a favourable variant which can take over and become the new wild type.

Long before this, an alternative approach had emerged offering a far more complex picture of the genetic structures of large populations (Provine, 1971). This derived from the study of continuous variation using statistical techniques originally pioneered by the biometrical school. Karl Pearson had regarded his model of variability in large populations as fundamentally incompatible with Mendelian heredity. He refused to allow any possibility of compromise along the lines suggested by Yule. But his student, Ronald Aylmer Fisher, eventually realized that Mendelism had come to stay and began to investigate how populations would behave if heredity was controlled by large numbers of Mendelian factors with overlapping effects (Box, 1978). Fisher's first paper (1918) was refused publication by the Royal Society of London, thanks to Pearson's influence, and eventually appeared in another journal. Fisher was able to show that natural selection would be effective in shifting the relative frequencies of genes in a population. His *Genetical Theory of Natural Selection* (1930) presented a sophisticated argument for gradual, adaptive evolution based on the mathematics of population genetics.

Instead of postulating a single wild type gene for each character, Fisher assumed that the population has a complex genetical structure in which many genes affect each character. Since mutated genes are often recessive, even unfavourable ones cannot be eliminated by natural selection simply killing off all the individuals in which the harmful character is expressed. Selection can only work to reduce the frequency of unfavourable genes, thus balancing the tendency for mutation to produce a constant flow of such characters. A character that begins to have a selective advantage – perhaps because of a change in the environment to which the population is exposed – will have its frequency slowly but steadily increased. Fisher showed that selection could act to maintain the balance between two alleles when the heterozygote is fitter than either homozygote. He developed an image of the population as a complex 'gene pool' within which the frequencies of individual genes can be altered by differential reproduction, i.e. natural selection. If the environment changed, selection did not have to wait for a favourable new mutation to appear, since the population already contained a wide range of genetic variability that could serve as the raw material of evolution. Although the effect of selection was slight in any one generation, if maintained over long periods of time the result would be significant evolutionary change.

Major contributions to the development of mathematical population genetics were also made by J. B. S. Haldane (1932; Clark, 1969). Like Fisher, Haldane calculated the effect of selection acting upon single genes within a large (theoretically infinite) population with random mating and perfect Mendelian segregation. But he used a practical example to show that the results of selection could sometimes be quite dramatic even within a short period of time. A dark or melanic form of the moth *Biston betularia* had first been noted in the early nineteenth century and had soon come to dominate the population in areas suffering from industrial pollution. Haldane assumed that the melanic form was a mutation whose takeover of the population had been made possible by the selective advantage it conferred as camouflage on soot-covered surfaces. He showed that so rapid a rate of change in the population implied a 50 per cent greater production of offspring by those individuals expressing the favoured gene, a far more intensive selective effect than Fisher had postulated.

Fisher and Haldane both adopted what Ernst Mayr (1959) has called the 'beanbag' approach to population genetics. They saw each mutated gene as an independent new factor fed into the gene pool to await the effects of selection. In fact geneticists were now recognizing that Mendelian factors are not always discrete units; often they interact so that the effect of two genes together is not the same as the sum of each acting on its own. In an attempt to show that the Mendelian factors themselves are variable, W. E. Castle (1911) had discovered that selection sometimes reveals additional variability beyond the normal range for the population. After a few years Castle was forced to give up the claim that the genes themselves are subject to variation, but his experiments had shown that genetic combinations can have unexpected effects. Selection in small populations could facilitate the appearance of unusual combinations of genes yielding phenotypic variation beyond the range normally observed. The amount of variation available as the raw material of natural selection would thus be greater if the population was not a homogeneously interbreeding mass of individuals, but was split into smaller groups where inbreeding could elicit these unusual genetic combinations.

Castle's student, Sewall Wright, used this interpretation of the population's genetic variability to construct a more complex view of the evolutionary process (1931; Provine, 1986). Using as his model the kind of inbreeding that occurs in small populations during artificial selection, Wright assumed that the wild population itself is seldom a single, fluid gene pool. He argued that in many cases the species would be broken up into relatively isolated local strains, with only limited interbreeding between them. In such small groups inbreeding would create new gene-interaction systems that would be seized upon by natural selection when they happened to confer some advantage. It has been widely assumed that Wright saw random 'genetic drift' as the source of nonadaptive characters in species, but Provine has shown that his real intention was to limit drift to the local groups. Only those resulting characters that are by chance adaptive will be able to spread out to colonize other sub-populations and eventually the whole species.

By introducing the possibility that local isolation could subdivide the population, Wright had produced a model of evolution that fitted more easily into the work of the field naturalists. Ernst Mayr (1976, 1982) argues that the study of biogeography and speciation had independently led field naturalists to concentrate on local adaptation as the cause of evolution, in effect endorsing the position outlined by Darwin. Originally inclined to view the adaptive process as Lamarckian, the naturalists had gradually become aware that the geneticists were undermining the plausibility of the inheritance of acquired characteristics and were now providing renewed support for natural selection. The theories of Fisher and Haldane were difficult to apply in practice because they ignored the role of geographical isolation. When Theodosius Dobzhansky (1937) began to turn Wright's complex mathematics into models of local adaptation that could be tested against field results, the stage was set for a synthesis that would establish a revived Darwinism as the dominant theory of evolution (Mayr and Provine, 1980). Naturalists such as Ernst Mayr (1942) and Julian Huxley (1942) proclaimed the superiority of the 'Modern Synthesis', while George Gaylord Simpson (1944) showed that the new approach was compatible with the data of palaeontology.

Up to this point, genetics had been alienated from the traditional biological disciplines, rejected as an upstart by field naturalists and palaeontologists alike. For this reason non-Darwinian evolutionary mechanisms such as Lamarckism and orthogenesis had remained popular within these traditional areas well into the 1930s. The synthetic form of Darwinism allowed the integration of genetics with the older disciplines. This in turn led to a final assault on the developmental view of inheritance and evolution which had survived the attacks of Darwinism and (in the traditional areas) of genetics itself. Evolutionists could no longer use the growth of the embryo as a model for the development of life on earth, and the recapitulation theory fell into disfavour. Changes in the growth process were now seen to be irrelevant to evolution unless those changes had been initiated by the appearance of new genetic factors. Lamarckism and orthogenesis by 'extensions to growth' became impossible. Evolution was now to be seen solely in terms of the genetics of populations, new factors being created by mutation and established in the population by natural selection when they conferred adaptive benefit.

The developmental viewpoint had exhibited considerable resilience, sustained by the widespread belief that the transmission of characters cannot be divorced from the process by which the characters are produced in the growing organism. The epigenetic view of development had held back acceptance of the view that germinal or genetic factors alone determine the essential character of the individual and hence the course of evolution. Now the evolutionists had joined the geneticists in accepting that only those aspects of growth that are predetermined by the genes are of relevance for future generations. The geneticists, in turn, had accepted that in the wild – as opposed to the artificial environment of the laboratory – adaptive consequences would determine which genes came to dominate the population. Genetics and Darwinism had emerged as parallel manifestations of the commitment to 'hard' heredity.

Opposition to Genetics

Orthodox histories of genetics and of evolution theory are likely to break off at this point, leaving their readers with an impression that classical gene theory and the Modern Synthesis have established themselves as permanent features of the scientific landscape. Subsequent developments have extended our knowledge of how the genes work, but have left the basic concepts intact. Only a handful of eccentrics have dared to challenge these firmly established theories, usually because they are motivated by outdated philosophical convictions. In reality the story is not as simple as this, and the very fact that the dissidents are branded as outsiders suggests that the triumph of genetics and Darwinism was partly the result of a social process going on within the scientific community. As with the reception of the original version of Mendelism, the degree of success enjoyed by classical genetics and revived Darwinism varied from one country to another, depending upon the beliefs and strategies adopted by the biologists who had the power to shape the way in which their science would be structured.

In America the Morgan school was immensely successful in creating research programmes and attracting funds. By defining heredity in its own strictly limited terms, it established itself at the centre of a new science capable of existing independently of traditional disciplines. We have already seen that long-established fields such as palaeontology refused to give up non-Mendelian and non-Darwinian ideas until the Modern Synthesis emerged in the 1940s. Embryologists also continued to insist that by turning their backs on development the geneticists had blinded themselves to problems that were insoluble in terms of the nuclear gene. E. G. Conklin and others argued that the extra-nuclear material of the cell, the cytoplasm, was crucial for development and hence, by implication, for inheritance. As far as these embryologists were concerned, the great debate over epigenesis and preformation had not yet been settled in favour of nuclear predetermination. Even Jacques Loeb, whose materialism had inspired Morgan, insisted (1916) that the organism is something more than a mosaic of genetic characters. Some more complex organization in the fertilized egg must control the developmental process, with many characters being influenced by the cytoplasm.

The litany of objections raised against classical genetics had a common source in the feeling that an atomistic approach to the problem of character-determination concealed the underlying unity of the organism. Whether materialist or vitalist, the opponents adopted a holistic or organismic view of the living system as something more than a collection of discrete elements (Haraway, 1976). The growth of so complex a system could not be predetermined by a string of independent genetic units in the chromosome; it was a co-ordinated process that must be controlled by a sophisticated organization in the cell as a whole. Many non-genetical biologists felt that the characters studied by the Morgan school were merely trivial or (in the case of mutations) harmful. The fundamental structure of the organism was determined outside the nucleus. As the geneticists became reconciled with Darwinism, their opponents began to insist that summing up individual mutations could never account for the evolution of complex adaptive structures. Some more purposeful process was required – either Lamarckism in which the acquired characters were inherited via the cytoplasm, or macromutations whose effect was co-ordinated by developmental factors. Palaeontologists frequently insisted that laboratory experiments could not be expected to reveal the presence of effects that would only have visible consequences after thousands of years. In various ways, these arguments can all be seen as efforts to preserve the developmental view of inheritance and evolution that had been shattered by the emergence of classical genetics.

Geneticists and Darwinists routinely dismiss these arguments as illustrating their opponents' commitment to an outdated philosophy of vitalism and teleology. When evidence is presented to demonstrate the existence of phenomena incompatible with the classical gene, it has been similarly dismissed as the product of wishful thinking if not downright fraud. The best-known examples of such attempts to pervert or obstruct the rise of modern biology are the 'case of the midwife toad' and the notorious episode centred on T. D. Lysenko's takeover of Soviet biology in the 1940s (discussed below). On the side of the geneticists who make these claims, it must be said that their opponents do tend to reflect a mode of biological thought whose origins lie in the developmentalist approach of the nineteenth century. The arguments often seem to attack only a caricature of Darwinism, a straw man designed to highlight the materialist implications of the orthodox view. The philosophical and ideological preconceptions of those who conduct their campaigns in this way are often all too obvious.

Yet the growing willingness of some modern biologists to question the complete adequacy of the genetical theory of natural selection alerts us to the fact that here – as is so often the case – history has been manipulated to give the establishment position an air of spurious authority. Instead of merely accepting the orthodox position, historians have recently begun to challenge the claim that the anti-genetic and anti-Darwinian arguments are merely politically or philosophically motivated attempts to sidetrack the natural course of biology's development. They have begun to suggest that the dominant paradigm of modern genetics and evolution theory is itself a product of professional and ideological decisions taken within the American and British scientific communities. As the status of the Modern Synthesis has come to seem less secure, some biologists have begun to admit that the opposing position may have identified real gaps in the orthodox framework of explanation. Historians in turn have asked different kinds of questions about how classical genetics and modern Darwinism came to dominate post-war biology in the English-speaking world. By looking more sympathetically at the positions adopted by the non-genetical biologists, it becomes possible to see that the success or failure of particular theories has not been determined solely by the available evidence. Non-scientific factors must be used to explain the success of the orthodox theory as well as the objections raised by dissidents. As in the case of Mendelism itself, one of the most powerful techniques used in this reassessment is a comparison of the ways in which classical genetics was received in different countries (Sapp, 1987). Far from enjoying universal success, classical and population genetics must be seen as theories whose rise to dominance was very much a characteristic of the English-speaking world.

In Britain, the chromosome theory of the gene gained little headway at first. Bateson was hostile because the notion of preformed genetic particles did not fit his own holistic view of living organization (Coleman, 1970). Mendelian experiments continued, of course, but Bateson's influence ensured that the chromosomal work of the *Drosophila* school did not catch on. In the 1920s Bateson at last conceded that there might be something of value in the chromosome theory, but it was a grudging admission qualified by a firm insistence that the problem of how the organism develops was not addressed by the new techniques. At the same time, however, Bateson was an unremitting opponent of Lamarckism who worked hard to establish the primacy of 'hard' heredity. His opponents in the biometrical school had an equally strong commitment to Galton's hereditarianism, a position which facilitated Fisher and Haldane's application of biometrical techniques to the genetics of populations. If at first only loosely connected with the chromosome theory, British genetics nevertheless upheld the view that the organism's genetic endowment is its only evolutionarily significant character.

We have already seen (Chapter 6) how the reception of Mendel-ism in France had been much more restrained. The new science gained no place in academic biology, due partly to the difficulty of establishing new disciplines in the highly centralized French system, and partly to a strong philosophical preference for non-mechanistic theories (Burian, Gayon and Zallen, 1988). Neo-Lamarckism continued to dominate French evolutionary thought (Boesiger, 1980; Limoges, 1980). Mendelian and chromosomally determined characters were widely dismissed as trivial, while embryologists stressed the role of the cytoplasm in controlling growth. Although useful work in population genetics was done by Georges Tessier and Phillipe L'Héritier, Tessier's elevation to a chair of genetics in Paris was largely a reward for his work in the resistance during the Second World War. Boris Ephrussi, holder of another post-war chair in genetics, was an embryologist by training. His influence ensured that the belated emergence of genetics in France would reflect interests quite different to those of the Morgan school. The study of cytoplasmic inheritance and of physiological genetics (the process by which the genetic material expresses its characters by influencing the growing organism) were seen as essential features of a genetics research programme. These interests helped to shape France's unique contributions to molecular biology in the post-war years.

Germany offers an even more instructive contrast with the English-speaking world, since here genetics flourished in the inter-war years, but did so in an environment which subordinated the study of transmission to broader concerns (Harwood, 1984, 1985; Sapp, 1987). Morgan's *Physical Basis of Heredity* was translated by Hans Nachtsheim, and Erwin Baur set up an active school devoted to exploring Mendelian and neo-Darwinian ideas. Yet elsewhere in the German system, genetics was treated as but one component of a comprehensive programme in the life sciences. As in France, the more centralized educational system made it difficult to establish an entirely new discipline. Most leading professors insisted on maintaining a traditional approach to inheri-tance in which transmission could not be separated from the problem of development. The relationship between heredity and evolution was of central concern, with palaeontologists looking to studies of heredity for ideas that would maintain the plausibility of Lamarckism and other non-Darwinian effects. Those biologists who accepted Mendelism and nuclear preformation tended to see these phenomena as only a limited part of the overall system of inheritance. They saw themselves as challenging the 'nuclear monopoly' established by the Americans. Even Carl Correns accepted the cytoplasm as a possible vehicle for the inheritance of acquired characters, while Wilhelm Johannsen conceded that the emergence of entirely new characters in evolution could not be explained by a series of genetic mutations.

Many German geneticists proposed a two-tier model of heredity and evolution, with Mendelian characters being responsible only for the trivial differences between species. Major evolutionary developments stemmed from changes in the *Grundstock*, a comprehensive system of inheritance encompassing the whole cell including the cytoplasm. Others adopted the *Plasmon* theory in which the whole cell was thought to be organized into a unified system with genetic elements in the cytoplasm co-ordinating the manifestation of all characters during growth. Most geneticists retained an interest in the growth process, although embryologists tended to be suspicious of all attempts to explain growth in preformationist terms. Hans Spemann promoted the concept of an 'organizer' field to co-ordinate growth and saw this field as residing in the cytoplasm (Horder and Weindling, 1985; Hamburger, 1988). Such ideas were widely hailed by neo-Lamarckians as offering possible mechanisms for the inheritance of acquired characters – although in some cases the geneticists were embarrassed by the use to which their theories were put by evolutionists.

The German enthusiasm for cytoplasmic inheritance and non-Darwinian evolutionary mechanisms was clearly at odds with the trend that became established in British and American biology during the inter-war years. This clash of cultures became obvious in those cases where German biologists actively confronted their English-speaking counterparts. Several German geneticists of Jewish extraction were forced to leave for America in the 1930s, where they found themselves unable to interact with the now-dominant Morgan school (Sapp, 1987). Victor Jollos had gained his reputation through the discovery of *Dauermodifikationen*, environmentally induced variations in micro-organisms which persisted for hundreds of generations before gradually fading away. He also studied heat-induced mutations, which he thought might be directed along a consistent path giving rise to orthogenetic evolution. Jollos believed that the cytoplasm played an important role in heredity, and his work was hailed by many anti-Darwinian evolutionists. When forced to leave for America in 1934, Jollos found his work was either ignored or actively criticized by supporters of the chromosome theory. He could not adjust to the different academic environment and soon found it impossible to get a job.

A more effective spokesman for the opposition was another refugee biologist, Richard Goldschmidt (Allen, 1974; Gilbert, 1988). Until forced to leave by the Nazis in 1935, Goldschmidt had been one of Germany's leading geneticists. From the start he was interested in physiological genetics and found it difficult to believe that the genes themselves could co-ordinate the complex process of embryonic growth. Although never a supporter of cytoplasmic inheritance, he suggested that the cytoplasm controls the timing of the process by which genetic characters are expressed. Goldschmidt also came to doubt the whole idea of the unit gene as conceived by the Morgan school. He argued that the position effect could be explained better by postulating the whole chromosome as the unit of inheritance. Mutations are not changes in individual genes, but rearrangements of chromosome segments which change the overall effect in a particular way. Turning to evolution theory, he argued that small 'mutations' could account only for minor changes. Significant evolutionary developments required macromutations – wholesale chromosomal rearrangements capable of producing entirely new characters. Goldschmidt coined the term 'hopeful monster' to highlight his belief that evolution could only advance through the appearance of monstrosities which occasionally conferred adaptive benefits. In America he was largely ignored by classical geneticists, but was able to get his views published. His *Material Basis of Evolution* of 1940 has become a classic source of anti-Darwinian arguments and has recently been reprinted with an introduction by Stephen Jay Gould (1980). Although Goldschmidt's saltationism has not gained wide support, he pinpointed weaknesses in classical genetics and paved the way for the emergence of a more sophisticated concept of the gene.

Although anti-Mendelian and anti-Darwinian arguments still flourished in the inter-war years, there was a shortage of hard evidence for the alternative mechanisms. Since genetics was founded on an experimental approach to heredity, its supporters were able to dismiss attacks on the adequacy of their conceptual scheme as mere philosophical quibbling. They also became expert in undermining the plausibility of the few reports of experimental evidence that the anti-Darwinians were able to bring forward. Arthur Koestler (1971) has drawn attention to the 'case of the midwife toad', in which evidence for Lamarckism supplied by the Austrian biologist Paul Kammerer was discredited under very strange circumstances. As far as the geneticists are concerned, this is a simple case of scientific fraud. But Koestler – a leading modern opponent of Darwinism – attempted to vindicate Kammerer by arguing that he had not been given a fair hearing. Few modern biologists would accept the implication that there might have been something in Kammerer's experiments after all, but Koestler's account does reveal that the geneticists' campaign against dissenting biologists was conducted with considerable ruthlessness.

Kammerer's original – and apparently successful – experiments to demonstrate the inheritance of acquired characters had been performed in the years before the First World War, using amphibians including the 'midwife toad'. In an effort to attract research funds in the difficult years after the war, Kammerer tried to interest British and American biologists in his work (1923, 1924). He also appealed to the popular press with calls for a new biology that would allow the breeding of an improved form of humanity. The media coverage alerted the geneticists to the threat posed by Kammerer's challenge, and Bateson became the leader of a campaign to discredit the experiments. Eventually one of the preserved specimens from the original experiments was shown to have been tampered with. Kammerer protested his innocence but committed suicide shortly afterwards, leaving the geneticists free to claim that his work had been a blatant fraud. The experiments were never duplicated, because Kammerer had an unrivalled ability to breed amphibians in captivity. It is this fact which allowed Koestler to claim that the geneticists, unable to disprove Kammerer's claims directly, had hounded him to despair with a campaign of innuendo.

At the time of his death, Kammerer was about to take up a position in the Soviet Union, where a unique scientific and political environment was allowing opposition to genetics to acquire its own momentum of intolerance. At first, genetics had flourished in Russia. Academic biologists such as N. I. Vavilov supported the Mendelian and chromosomal theories, while population genetics received a major impetus under C. C. Chetverikov (Adams, 1968). But from the start an alternative view of inheritance was promoted in the agricultural institutions. Experimental work was performed by T. D. Lysenko and his followers to confirm the inheritance of acquired characters, especially through the 'vernalization' of wheat (producing faster-growing varieties by freezing the seeds). In the 1930s Lysenko began to argue more actively against genetics, adopting an increasingly political tone in which the Morgan school was branded as an offshoot of idealism and hence incompatible with Marxist philosophy. By the 1940s Lysenko had gained the support of 'Stalin with claims that his approach would revitalize Soviet agriculture. He engineered a takeover of the scientific community and drove geneticists such as Vavilov from their positions. Lysenko retained control of Russian biology until the Khrushchev era, in what is popularly seen as a sinister attempt to pervert the course of science by the application of political pressure (Joravsky, 1970; Medvedev, 1969).

Marxist biologists accept that science and ideology cannot be separated, but concede that the Lysenko affair illustrates the dangers that may arise when the relationship is mishandled (Lewontin and Levins, 1976; Lecourt, 1977). More recently, historians of biology have begun to outline a more balanced interpretation based on a recognition that classical genetics cannot be seen as an ideologically neutral response to the problem of heredity in the 1930s (Roll-Hansen, 1985; Sapp, 1987). Genetics triumphed in America and Britain by establishing itself as a new science with an artificial focus on the problem of transmission, and by discouraging research in areas that might threaten the claim that the nuclear gene was the sole determinant of heredity. Lysenko's approach seemed outlandish by the standards of classical genetics, but it arose from a consideration of issues that were still taken seriously by most biologists outside the English-speaking world. Far from being a complete charlatan, Lysenko's early physiological work was quite sound. Even his ideas on heredity were compatible with the broader concerns of French and German biologists, and he was able to call upon their work to support his claims. Although eventually perverted by Marxist rhetoric, Lysenko's views articulated concerns that were shared by many biologists who suspected that classical genetics had cut itself off from issues that should have remained central to biological research.

No one doubts that Lysenko misused his political power when he eliminated all support for nuclear preformation. But if his position represents a dogmatism of the left, is it not possible that classical genetics was itself shaped in part by ideological pressures? In order to sustain their 'nuclear monopoly' the English-speaking geneticists had branded the study of certain topics as heretical, even though those topics were still seen as central in importance by continental biologists. The fate of Jollos and Kammerer indicates that the paradigm's supporters could be viciously intolerant of dissent. In the post-war era genetics became caught up in the cold war: to study cytoplasmic inheritance was to support Lysenko and hence to side with communism (Sapp, 1987). Tracy Sonneborn, an American biologists who worked on cytoplasmic effects, found himself branded as a Lysenkoist. Western geneticists might write of *The Death of a Science in Russia* (Zirkle, 1949), but they were unable to see that genetics itself was an artificial construct of American and British science. The Lysenko affair can no longer be judged in black and white terms, and it forces us to confront the possibility that there may be an ideological dimension to the rise of hereditarian theories.

8

Heredity and Politics

We have already seen that the emergence of modern genetics can be explained partly in terms of political manoeuvering within the scientific community. The early Mendelians and the classical geneticists made a bid for advanced status within the framework of academic biology. To do this they redefined the concept of heredity to focus attention on the problems that they alone had the techniques to solve. This in turn allowed them to attract research funds from bodies concerned with agricultural breeding and to set up autonomous programmes in universities and breeding stations. In countries such as France and Germany, where the academic system made it difficult to set up new units, this technique was unsuitable and the study of heredity remained subordinated to more traditional biological problems.

The Lysenko affair shows that debates over heredity also interacted with politics in the normal sense of the term. Lamarckism and cytoplasmic inheritance became identified with support for the Soviet system, while classical genetics was defined as 'pure' science only in Britain and America. In fact, the rise of hereditarian theories had an ideological dimension from the start. A theory of biological inheritance inevitably carried implications for the inheritance of human characteristics, thereby impinging on philosophical and social concerns. By the end of the nineteenth century, the philosophy of scientific naturalism had popularized the view that a person's character was determined by the physical structure of his or her brain – not by a spiritual entity, the soul. Human nature itself thus became a potential subject for control by scientific manipulation of the individual's physical characteristics. The critical question was: how best to take control? Which kind of biological process offered the best chance of understanding and eventually controlling the human population? Hereditarian theories can be seen as an answer to this question. If heredity played the most important role in determining human characters, then only a better knowledge of heredity would allow social control to become a reality.

Hereditarian theories represent one extreme in the great debate over whether nature or nurture determines human character (Pastore, 1949; Cravens, 1978). 'Nature' in this context means genetic inheritance: a person's character is predetermined by his or her genes and can only be modified to a trivial extent by environmental factors and education. 'Nurture' refers of course to these environmental factors. If nurture is more important than nature, a person's character is not predetermined by heredity and can be shaped to a significant extent by the way he or she is raised and educated. Hereditarian theories can be seen as the biologists' vehicle for promoting the claim that they – and not the educators – have the key to understanding how human nature may be controlled. If this claim is accepted, improvement of the population can only be achieved by using knowledge of heredity to control human breeding, ensuring that individuals bearing superior characters reproduce more rapidly than those less fortunately endowed by biology. Francis Galton coined the term 'eugenics' to denote this kind of artificial breeding programme for the human race. Such a policy enjoyed considerable support in the early decades of the twentieth century and reached its apogee in the Nazis' efforts to 'purify' the Aryan 'race'.

The all too obvious excesses of the Nazis helped to stem the tide of support for the more limited means proposed in other countries to limit the breeding of the 'unfit'. And yet the claim that heredity determines character continues to be used as a means for urging less blatant measures of social control. This is perhaps most apparent in the field of education, where it is sometimes claimed that IQ (intelligence quotient) is determined by heredity, and that if a good education can only benefit those who have inherited the genes for high intelligence, it is a waste of public money to apply the same level of education to the less well-endowed. In the United Kingdom, this philosophy underpinned the two-tier educational system which existed until quite recently (and which still survives in Northern Ireland). Critics maintain that the policy is designed to benefit the upper and middle classes by condemning the working class to a permanently inferior status imposed by education. In America, similar controversies have raged around the claim that certain racial groups have an inferior IQ, and hence a limited educational potential. Class and race are the twin centres around which the debate over biological determinism revolves. A third potential centre is sex, although this has received less attention so far. Even so, the possibility that the notion of genes determining character reflects a distinctly masculine view of things has been hinted at by biologists such as Barbara McClintock (Keller, 1983).

Because the supporters of biological determinism frequently identify particular races or social classes as the source of 'unfit' characters, the whole philosophy is sometimes branded as a reflection of right-wing ideology. There is something in this generalization, but the link with the right is not as clear-cut as opponents would like us to believe. The population geneticist J. B. S. Haldane wrote openly *against* eugenics (I have borrowed the title of his book (1938) for this chapter). Hermann J. Muller, one of the architects of classical genetics, shared Haldane's socialist perspective and spent some years in Soviet Russia – yet he actively supported the idea of a breeding programme to improve the human race (Carlson, 1981). Arguments based on the strength of heredity have often been used to justify a hierarchically ranked society, but the political left has historical links with eugenics which its modern supporters would prefer us to forget (Paul, 1984).

If there is one thing upon which politicians of the extreme left and the extreme right agree, it is a distrust of liberal, *laissez-faire* policies. Both sides want to *control* society, and although they disagree over the goals to be achieved they may sometimes endorse similar mechanisms. The right prefers biological determinism because this builds the notion of a social hierarchy into the system from the beginning: human beings are unequal because they inherit unequal genetic potential. The left wants to dismantle the hierarchy and has sometimes conceded that it would be easier to do this if biologically inferior characters were eliminated. In general, however, the left has inclined to the opposite side of the nature–nurture debate. It has argued that a better society can be produced by using equality of upbringing and education to eliminate the differences in ability that the determinists attribute to inheritance. Sociologists too have taken the side of nurture – not surprisingly, since it is in their interests to claim that social rather than biological measures are needed to shape the future of humanity.

Biologists who accept that individual development is shaped by environmental factors, not by a preformed germ plasm, leave room for nurture in the shaping of human character. Epigenetic theories of growth certainly fall into this category, and we have seen that such theories were frequently associated with a belief in the inheritance of acquired characters. There is a long-standing association between Lamarckism and calls for social reform. New characters can only be acquired if the organism is not rigidly determined by heredity – yet the possibility that the acquired characters themselves can become inherited offers the prospect that improvements produced by a better environment will become of permanent value to the race. Lamarckism thus blurs the distinction between nature and nurture, generally to the benefit of nurture. Lest Lamarckism gain an untarnished image as the reformers' biology, however, it must be pointed out that the theory played a major role in establishing nineteenth-century views of racial inequality.

What role did these social implications play in the debates that led to the creation of modern genetics? Historians who adopt the modern fashion of analysing theoretical debates in terms of their social consequences (see Chapter 1) find it difficult to believe that the rise of hereditarian theories was unconnected with the parallel debates on the determination of human characters. The link is certainly not one in which science itself is rigidly determined by ideological factors. No one is claiming that genetics emerged because it was the only theory capable of satisfying the popular demand for a model of society based on biological determinism. In fact, a whole range of theories made bids to serve in this function, although genetics was certainly the major beneficiary. The belief that social problems were caused by an ever-increasing proportion of biologically inferior people inevitably focused attention upon the debate over biological inheritance. At the very least, the innovations which led to the establishment of genetics were thrown into unnatural prominence by the widespread assumption that the topic had become of pressing social importance. To those who took the side of nature (heredity), any theory which promised to substantiate their feeling that characters are rigidly predetermined at birth was welcome. By focusing attention so firmly on transmission, Mendelism satisfied this demand. There is no doubt that the theory's prospects – in some quarters, at least – were enhanced by its social implications. The extent to which the theoretical innovations proposed by individual biologists were shaped by this aspect of their cultural environment remains a fertile source of debate among historians.

Social Darwinism

There is a widespread assumption that the philosophy of biological determinism first rose to prominence in the wake of Darwin's theory of natural selection. This theory certainly implies that unfit (i.e. maladaptive) characters must be constantly eliminated for evolution to take place. It would thus have been possible for anyone at the time to draw a connection between biological evolution and social progress, arguing that the latter would only occur if unfit members of the human race were weeded out. In the interpretation popularized by Richard Hofstadter (1955), such a policy of 'social Darwinism' was a prominent theme in late-nineteenth-century political thought. Supporters of the free-enterprise system drew on the theory of the 'survival of the fittest' to argue that it was natural for society to advance by allowing the unfit to be eliminated by the consequences of their own inability to cope with the world. For Hofstadter, eugenics was merely a continuation of social Darwinism by other means, a switch from natural to artificial means of limiting the reproduction of the unfit. Harsh social policies based on the assumption of biological inequality are thus to be seen as direct continuations of the view of human nature first popularized by Darwinism. For the modern critics of biological determinism, the claim that each new manifestation of the hated philosophy is no more than a revival of social Darwinism has become an integral part of their debating rhetoric.

Before moving on to describe the heyday of hereditarian thought, I want to argue that the policy of tracing all forms of biological determinism back to the era of Darwinism is misguided. Such an interpretation presents a distorted image of nineteenth-century social Darwinism and conceals the extent to which the use of biology in social debates has changed over the intervening years. In particular, it conceals an important change of public attitude which coincided with the emergence of Mendelism as the dominant theory of hard heredity. For all that it is called 'social Darwinism', the use of evolution theory to support the free-enterprise system by thinkers such as Herbert Spencer was not a manifestation of the hereditarian aspects of Darwin's theory that have been adopted by modern biologists. The genetical theory of natural selection highlights aspects of Darwin's thinking that were only imperfectly appreciated by his contemporaries, who preferred to see evolution as a purposeful system designed to encourage the whole population to greater efforts.

Social Darwinism took progress for granted and was, in this respect, a product of the nineteenth century's fascination with growth as the model for all natural development. Its emphasis on struggle as the motor of progress did not follow the logic of Darwin's mechanism, but saw competition as a stimulus that would encourage all to greater achievements. The purpose of struggle was not primarily to eliminate the congenitally unfit, but to encourage the whole population to become fitter. Far from advocating a view of human nature as rigidly determined by heredity, such a policy depended upon the belief that individuals can better themselves if suitably encouraged. What the modern biologist sees as the key point of Darwin's theory was thus subverted in that theory's most popular social application. Social Darwinism was really social Lamarckism, a reflection of the pre-hereditarian view of human nature. Only in the field of race relations, where the 'inferior' races were supposed to be fixed in their character through centuries of exposure to poor conditions, did the nineteenth-century view of human nature anticipate the rigid hereditarianism of the era of Mendelism. If this interpretation is accepted, the emergence of hereditarian theories and social policies around 1900 represents a major break with the past and a transformation of the way in which biology was used to provide a model for social policies.

In Hofstadter's interpretation, it was the philosopher Herbert Spencer who transmitted the gospel of social Darwinism from Britain to America. Spencer was certainly a leading advocate of the *laissez-faire* approach to social policy (1851, 1884; Peel, 1971). Each individual was expected to look after his own interests, and must face the consequences of any failure to cope with the social environment. Since it was Spencer who coined the term 'survival of the fittest' to describe the effects of natural selection, it seems obvious at first sight that his intention must have been to promote a model of society based on Darwinian principles. Social and economic progress would be guaranteed by allowing free-for-all struggle in which any individual with a congenital weakness will be eliminated by starvation, leaving the 'fittest' free to exercise their abilities to the fullest. Spencer's philosophy of extreme individualism flourished in late-nineteenth-century America, where the 'robber barons' of industry were only too willing to quote his views in endorsement of their ruthless attitude to business.

There are many problems with this interpretation, however, not the least being that 'social Darwinism' was almost always used as a term of abuse (Bannister, 1979). No one admitted to being a social Darwinist, but many were labelled thus by enemies seeking to highlight the ruthlessness of their methods. The image of an age dominated by Darwinian principles is to some extent an artefact of liberal historians seeking to emphasize the extent to which modern society has progressed beyond such unpleasantness. Although many late-nineteenth-century thinkers used Darwinian metaphors to support their political beliefs, there was no single version of 'social Darwinism'. Politicians of all persuasions sought to associate their views with scientific evolutionism as a means of gaining respectability (Jones, 1980). In many cases the links thus established were based on a complete misunderstanding of Darwin's theory.

Spencer's individualism had already been worked out before the *Origin of Species* was published, and – despite his coining of the term 'survival of the fittest' – Spencer remained a convinced Lamarckian. He acknowledged that in extreme cases those individuals who are unable to cope with social evolution will be eliminated. But this was only a secondary, purely negative effect. The real purpose of individual competition was to encourage everyone to respond in the most active way to the challenge of an evolving society. People are forced to adapt as best they can, knowing full well that there is no cushion provided by the state to soften the misery that is the consequence of failure. Fear of failure thus stimulates initiative: people learn from their mistakes because they are punished so forcefully that they become determined to do better in future. They also teach their children the benefits of their hard-won experience, thus ensuring that the habit of adaptability becomes ever more deeply engrained in the population. Here Spencer's Lamarckism comes into play, allowing the species as well as the individual to learn the lesson taught by exposure to a changing environment.

It was this emphasis on the stimulating effects of competition that was taken up by the American businessmen. Whatever the harsh consequences of their policies in practice, in principle they were unwilling to condone the weeding out of all those with less than average abilities. Everyone had a chance to rise in society, and the extent of the rise would be determined by the initiative shown in response to the challenge. As Moore (1985) points out, many of Spencer's followers were committed to a Christianized version of evolutionary progress. Their 'social Darwinism' was merely the Protestant work ethic in a new guise, with the rewards for effort and enterprise being given in this world rather than the next.

Far from placing the emphasis on congenital limits to individual ability, Spencer's philosophy stressed that almost everyone would respond positively when challenged by the environment. The free enterprise system is the best way of forcing people to learn from experience. Spencer believed that society was too complex for us to take charge in an attempt to direct its progress artificially. Any attempt by governments to interfere, however well intentioned, would backfire and do more harm than good. Yet later in the century the Lamarckian view of human nature was also used to support the increasingly popular view that governments *should* take charge of social development (Stocking, 1962, 1968). Sociologists such as Lester Frank Ward argued that we now have the ability to control social and ultimately biological evolution. If we institute social reform, people will be improved by exposure to better conditions, and the benefits will be inherited by their children to produce a permanent effect on the human character. Education offers the possibility of controlling how our children behave, with the inheritance of acquired characters ensuring that the new behaviour-patterns are transmitted to future generations as instincts that have become incorporated into human nature.

This approach is sometimes called 'reform Darwinism', but in fact it was another application of Lamarckism to human affairs. Deliberate choice rather than natural evolution was to direct the future of the race by controlling the acquisition of new characters. As late as the 1920s Paul Kammerer, the biologist at the centre of the 'case of the midwife toad', attracted considerable publicity by claiming that Lamarckism would allow the human race to be improved (Koestler, 1971). By this time, however, the geneticists had begun to resist all efforts to show that character is shaped by the environment rather than by inherited genes. Kammerer was one of the last advocates of what might be called 'social Lamarckism', although reformers have continued to argue that better conditions produce better people, even if the effects are not inherited.

Whatever the hereditarian implications of Darwin's selection theory, it can be argued that the chief applications of evolutionism to social affairs did not take up this aspect of his mechanism. As in biology itself, the *Origin of Species* popularized evolutionism, but could not displace a Lamarckian and developmental view of human nature (Bowler, 1988). Few were as yet prepared to accept that a person's abilities are rigidly limited by inheritance. There was, however, one exception to this: virtually all nineteenth-century thinkers were prepared to argue that the various races of mankind exhibit different levels of mental ability. This was an explicitly hereditarian approach, although it stood at variance with the general belief that in the white race, at least, individuals could be stimulated to greater efforts by an external challenge. In their expansion around the globe, Europeans had become convinced that their technological sophistication was a sign of intellectual superiority over other races (J. S. Haller, 1975; Stepan, 1982). Some race theorists had even proposed that the human races were separately created species. Evolutionism obviously undermined this position, but many evolutionists remained convinced that the races were the products of parallel lines of human development which had remained distinct over vast periods of time (Bowler, 1986). Whatever the mechanism of adaptation, it was thus possible to argue that the Europeans' long exposure to a more challenging northern environment has gradually endowed them with a permanent superiority over races evolved in softer climes.

Even Lamarckians such as Spencer were anxious to support the view that the non-white races had by now acquired a permanently inferior mentality. An improved environment might, in theory, allow them to advance further up the scale of mental evolution, but in practice the effect would be so slow that there was no hope of them ever catching up with the triumphant whites. The black and brown races were, in effect, condemned to permanent inferiority by the sheer length of time in which they had been exposed to less stimulating conditions. Lamarckism thus became a hereditarian philosophy in the area of race. Its link with the recapitulation theory and the analogy between evolution and growth also allowed it to support a hierarchical ranking of the races. The American neo-Lamarckians in particular argued that the non-white races occupied lower rungs on the ladder of development – they were relics of the ancestral stages through which the whites had evolved, permanently stuck at a lower point on the scale because of their exposure to a less stimulating environment. The claim that the coloured races preserved the ape-like characters of the white's prehuman ancestors exploited the developmental model of evolution to reinforce the sense of racial superiority already imprinted on the European mind.

Race thus provided a model for the subsequent development of hereditarian ideologies. Even those nineteenth-century thinkers who accepted the power of the individual to respond in a positive way to environmental stimulus refused to extent this quality to the non-white races. A few generations of reform could not hope to undo the deleterious effects of millenia of evolution in a poor environment. These races were condemned to permanent in-feriority by the inherited consequences of their evolutionary past. The application of a similar policy of biological determinism within the white race itself would become possible only when it could be argued that the individual's character is rigidly limited by the inherited effects of family as well as racial background. Then it would become possible to argue that social class is also a determinant of character and ability.

Eugenics

To be consistent with the theory of natural selection, a social Darwinist would have to argue that (*a*) there is congenital variation in the level of intellect and ability between individual human beings, and (*b*) that the best way of ensuring progress is to follow nature by allowing the least fit to be eliminated by starvation. In practice, everyone (Spencer included) accepted that civilized values were not consistent with such a bleak policy: the weak were inevitably protected, if only by individual acts of charity. Darwin himself realized that our tendency to help the less effective members of society will lower the overall standard of the human race. If a return to nature was impossible, the only alternative for a hereditarian thinker would be to advocate a policy of artificial selection for mankind. Since some people are born with a lower than average level of ability, and will transmit their inferiority to their children, society must take steps to limit the proportion of such individuals in the population. The state must somehow limit the reproduction of the unfit, while encouraging the more intelligent to increase the size of their families. Under the name 'eugenics', such policies became the most active expression of hereditarian thought in the early twentieth century. The unfit – usually identified as members of the lowest social class – were to be forcibly prevented from breeding. Members of lower races were to be forbidden entry into countries populated by Europeans, lest their inferior characteristics contaminate the race.

The originator of the term 'eugenics' was Darwin's cousin, Francis Galton (Forrest, 1974). In his *Hereditary Genius* of 1869, Galton advanced the basic hereditarian position. He claimed that the children of eminent fathers tend to inherit their superior qualities: they are born with above-average ability, while the children of inferior parents are condemned by heredity to a lower status. Galton was already convinced that society must ensure the more active reproduction of its fitter members if it is not to be swamped by the unfit. Significantly, his views were at first repudiated by the majority of his contemporaries, suggesting that the hereditarian component of Darwinism had not yet been appreciated. Galton's continuing studies of heredity were, to a large extent, undertaken in an attempt to vindicate the position he had sketched in (Cowan, 1972b, 1977).

Eugenics became a kind of moral crusade into which Galton threw most of his energy. He was determined to alert society to the danger of allowing the unrestrained reproduction of the unfit, and to the positive benefits that could be gained from taking control of human breeding. At the bottom end of the spectrum of abilities, negative eugenics required that the feeble-minded should be prevented from reproducing either by institutionalization or by actual sterilization. More positively, the professional class (where Galton believed the highest levels of ability were concentrated) should be encouraged by tax incentives to have more children. By the 1890s Galton had been joined by his student, Karl Pearson, and their campaign began to win support. Pearson (1901) noted that recruitment for the army in the Boer War had exposed the poor mental and physical state of many working-class men. The slums were now portrayed as breeding grounds for a mass of substandard humanity, an ever-growing strain on the nation's resources and a potential threat to its security. In 1904 Galton founded the National Eugenics Laboratory, followed soon afterwards by the Eugenics Education Society and the *Eugenics Review*. The movement's great triumph came in 1913 with the passing of a Mental Deficiency Act requiring the institutionalization of the severely retarded. (On eugenics in Britain see Farrall, 1979; Mackenzie, 1982; Searle, 1976, 1979; Kevles, 1985; Jones, 1986).

In America, eugenics also began to attract widespread support (Haller, 1963; Pickens, 1968; Ludmerer, 1972; Allen, 1976; Kevles, 1985). The American Breeders' Association set up a Eugenics Committee in 1906 and in 1910 the Eugenics Records Office was founded to co-ordinate research (Allen, 1986b). If anything, the American approach to eugenics was harsher than its British counterpart. A number of states passed legislation requiring the sterilization of individuals found to be below a certain standard of intelligence. The movement also became associated with the campaign to restrict the immigration of 'inferior' racial types that might threaten the whites' domination of the population. A flood of literature proclaimed the danger of allowing the fast-breeding yellow and brown races, or even the eastern Europeans, to gain a foothold. In 1924 the Immigration Restriction Act ensured that all potential immigrants would be tested, and from this date onwards many were, in fact, turned back. The effects of this Act were to become particularly unfortunate after the rise of the Nazis in Germany, since their campaign for racial purity led to a an exodus of Jews and other 'undesirables'. The Nazis had their own eugenics programme, although this merely built upon foundations already established in early-twentieth-century Germany (Weiss, 1987).

To some extent eugenics was a product of the wave of enthusiasm for social management that replaced the *laissez-faire* attitude of the Victorian era. Science was to be used as a means of social control, allowing governments to understand and direct the forces at work within the population. But which sciences were the most relevant? If one accepted that nurture was more important than nature, then biology had little relevance because the social factors that shape a person's upbringing offer the most effective way of influencing his input into society. The newly emerging social sciences thus had a vested interest in promoting this solution to the problem, since they would supply the expertise that would direct the reforms designed to improve the character of the lower classes. For this reason, social scientists were and still are inveterate opponents of the hereditarian position (Cravens, 1978).

The only problem with reforms intended to help the lower classes was that they were expensive. To members of the newly-emerged professional class, the thought that their hard-earned salaries might be taxed at an ever-higher rate to benefit the lower orders was deeply worrying. By emphasizing nature rather than nurture they could argue that reform was a waste of time and money: the lower classes were fixed at an inferior level of ability and no amount of improved health-care or education could benefit them. Hereditarianism seemed to offer a cheaper solution to the growing problems of society, since instead of throwing money away on reform one could simply eliminate the worse elements by preventing them from breeding. From this perspective, biology went straight to the root of society's ills and offered the only valid approach to social management. The study of heredity thus became the essential source of information on how inferior characters are maintained in the population and on how they might be eliminated (Paul, 1985). Psychology also benefited from this approach, since it could provide intelligence tests that would identify those individuals who had inherited a defective mental character. Many of the early IQ tests were, in fact, grossly inadequate (Gould, 1981; Evans and Waites, 1981). People were labelled as being of low intelligence because of a lack of factual knowledge that would now be attributed to poor education. Controversy still surrounds the work of Cyril Burt, a psychologist involved in the setting-up of Britain's two-tier educational system (Hearnshaw, 1979). Burt's tests on identical twins raised in different environments were supposed to show that intelligence is rigidly determined by heredity, but some scholars claim that his results were falsified.

But what effect did this movement have on the study of heredity? At the very least, it seems plausible to believe that the intense excitement surrounding this field of research at the turn of the century was aroused by the knowledge that it had important political implications. Eugenics created a demand for theories which stressed the role of heredity over environment in the shaping of human character, and it can surely be no accident that the concept of pure heredity was clarified at this time. Hereditarian theories in biology were to some extent a product of the newly-emerging hereditarian attitude in Western culture. It is necessary to qualify this statement, however, by insisting that we cannot treat Mendelism and classical genetics as 'nothing but' a product of imaginations disordered by ideological pressures. Not every gene-ticist was a supporter of eugenics, and there were rival theoretical approaches capable of satisfying the hereditarians' demand for a scientific foundation to justify their prejudices. The exact course of biology's development in the early twentieth century was shaped both by ideological pressures and by forces at work in the national scientific communities, as we have seen in earlier chapters. And, of course, the available evidence had to be taken into account – although we have seen that theoretical preconceptions can determine a scientist's view of what evidence is relevant to his or her particular concerns.

The fact that hereditarian social values need not generate a belief in particulate inheritance is obvious from Galton's original ap-proach to the question. He decided to verify his belief in the importance of heredity by studying the transmission of continuously varying characters within the whole population, not by trying to isolate unit characters in experimental samples. Galton's support for saltative evolution suggests that his eugenic beliefs were not a form of 'social Darwinism', but Pearson clearly visualized the elimination of socially undesirable characters as an aritificial equivalent of natural selection. His eugenic programme was based on a theoretical foundation that was positively hostile to the ideas of saltative evolution and particulate inheritance. Although strongly committed to hard heredity, he repudiated Bateson's Mendelism and promoted a non-genetic justification of eugenic policies. Mackenzie (1982) argues that the structure of Pearson's statistical techniques was designed to substantiate his hereditarian beliefs. A little later on, R. A. Fisher's efforts to link biometry and Mendelism were almost certainly inspired by his strong support for eugenics (Box, 1978; Bennett, 1983). Fisher's *Genetical Theory of Natural Selection* (1930) contains extensive discussions of heredity in the human population. The possibility that mental and physical defects were transmitted by Mendelian factors became a central theme of eugenics in the 1920s and 30s.

Yet the transition from a biometrical to a Mendelian justification for eugenics was by no means smooth. Pearson refused to accept that Mendelism could throw light on the distribution of inferior characters in the population as a whole. Conversely, Bateson refused to associate his advocacy of Mendelism with eugenics. Apart from professional hostility to Pearson, his attitude may have been conditioned by a more conservative élitism which led him to dismiss the social management approach as a product of bourgeois thinking (Mackenzie, 1982). Bateson accepted that heredity was important, but refused to endorse the political application of this knowledge. J. B. S. Haldane (1938) resisted eugenics from a socialist perspective, conceding the genetic basis of some defects but pointing out that the human population is so complex that artificial selection has no realistic hope of eliminating the responsible genes within a few generations. One of Haldane's chief targets was the embryologist E. W. MacBride, who advocated a harsh regime of sterilization for the unfit from a Lamarckian perspective (Bowler, 1984b). MacBride was able to use Lamarckism as a foundation for hereditarian policies by developing a link between class and race: he claimed that the worst elements in British society were of Irish descent, carrying with them the legacy of millenia of evolution in a poor environment. The fact that a Lamarckian could support eugenics while some geneticists did not provides clear evidence against the claim that the development of biology was shaped solely by ideological pressures. There *was* a demand for hereditarian ideas to support eugenics, and sympathetic biologists from various backgrounds tried to satisfy that demand. In the end Mendelism emerged as the most effective way of trying to show that defective characters have their origins in heredity, but it achieved this status only after considerable debate.

Eugenics did not create genetics, but it did create a climate of opinion in which this kind of hereditarian theory could flourish. Mendelism achieved its greatest success in America, and it is perhaps not coincidental that here its association with eugenics became apparent at an earlier date. From the start there were American geneticists willing to develop the theme that the transmission of defective characters in the human population could be explained in terms of harmful mutations perpetuated according to Mendel's laws. C. B. Davenport in particular became notorious for his efforts to identify a genetic basis for all mental and physical defects. He was even prepared to argue that feeble-mindedness was due to a single recessive gene. In the context of the early geneticists' efforts to identify unit characters that could be traced with their techniques, this claim seemed all too plausible and served as the inspiration for a good deal of early genetic research. At a more sophisticated level, H. J. Muller was aware of the potential implications of his work for the human race and – despite his socialist leanings – advocated the improvement of the race by artificial insemination with the semen of superior men (1935). Other geneticists such as T. H. Morgan and Sewall Wright remained aloof, however. They knew that genetics had gained some of its early support because of its appeal to hereditarian social thinkers, but they did not think it appropriate for biologists to allow this to direct their research.

These suspicions became even more pronounced as the geneticists realized that popular support for eugenics was being whipped up by appeals to grossly oversimplified versions of their science. Such appeals not only ignored the possibility of environmental effects on human growth, but also adopted an oversimplified view of the relationship between genes and characters. The possibility that intelligence might be influenced by a number of interacting genes was ignored because it did not fit the eugenists' hopes that a small number of defective genes might be eliminated by a simple process of selection. By the 1920s, however, genetics had begun to reveal a far more complex picture in which it was impossible to identify a single gene as responsible for so diffuse an effect as 'low intelligence'. Population genetics was also beginning to show how difficult it would be for artificial selection to remove defective genes that had become established in the human gene pool. In the 1930s the excesses of the Nazis' eugenic policy began to generate widespread public concern in other countries. What had begun as a campaign to limit the spread of defective characters could now be seen as a policy that could all too easily be exploited by groups with a preconceived idea of what the 'best' form of humanity should be. In its original form, at least, eugenics began to slip into the background, with even supporters conceding that nothing could be done until better conditions for the poor allowed a reasonable assessment of their genetic potential (Bajema, 1977; Kevles, 1985).

In the post-war era eugenics as such has seldom been openly advocated, but the claim that human character is to a significant extent determined by heredity continues to emerge in debates over social policy. The controversy surrounding Cyril Burt's allegedly fraudulent support for hereditarianism in educational policy indicates that the issue is still a sensitive one. In America there have been attempts to argue that the black and white races have different intellectual potentials, preserved by heredity. The most controversial advocate of this view has been Arthur S. Jensen (1969, 1972). At the same time the emergence of sociobiology has generated the claim that some human behavioural patterns may be imprinted on our brains as the result of evolution (Wilson, 1975, 1978). Although originating in the modern revival of Darwinism, sociobiology depends for its effect upon the belief that character is determined by genes rather than by upbringing. Its claims have been vigorously repudiated by social scientists and liberal thinkers committed to the view that reform is the key to improving the character of the human race (Caplan, 1978). Biologists too have sought to resist the various attempts to retain a role for genetic determinism (Rose, Kamin and Lewontin, 1984).

Modern biological determinism has many roots, but the classical genetics of the 1930s can hardly be seen as a continuing influence. The belief that heredity does have a major role to play in determining human character may well have been boosted by the post-war developments in molecular biology, which have – so far as the general public is concerned – 'solved' the problem of how characters are transmitted and expressed (Kaye, 1986). The discovery of the structure and function of DNA is often presented as the key that has unlocked the most fundamental secrets of life itself. Small wonder that nonspecialists reading such accounts of the latest developments in molecular biology are encouraged to believe that heredity must, after all, be the foundation of organic structure and hence of human character. Eugenics is dead, but biological determinism continues because the transition from classical to molecular genetics has been presented as the great success-story of modern biology. Heredity remains central to the public's perception of biology's social role, and it does so because in transcending the limits of classical genetics the biologists have opened up a new world of biotechnology. If they can control life at will, even manufacturing new organisms to required specifications, it is because they can direct the process that generates living structures from a molecular code. Emphasis on the genetic code as a complete representation of the organism itself helps to fuel the popular belief that human nature too is predetermined by our genes.

9

Epilogue: from Classical to Molecular Genetics

Classical genetics developed and applied the theory of the gene as a determinant of the organism's physical, and to some extent mental, characteristics. Yet to a large extent it did not address the question of how the gene encoded the information it transmitted, nor of how that information was expressed in the development of the organism. By itself the basic idea of a 'unit' of heredity located on the chromosome proved remarkably fruitful in terms of both pure research and practical application to breeding and agriculture. The decades following the Second World War have seen major developments that might almost be counted as a second revolution in genetics. Molecular biology has allowed us both to ask and to answer those questions about the nature of the gene and its means of expression that the classical geneticists were forced to ignore. To complete our survey of the Mendelian revolution we must sketch in the events of this second phase, showing how modern theories and techniques have extended and to some extent modified the conceptual structure of classical genetics. We must also assess the impact that modern theories have had upon our broader conception of life, including human life. Since the new techniques have opened up a Pandora's box of practical applications, it is important to ask if our views about the nature and value of life will be affected by our ability to control its development.

At one level, the new theories can be seen as an extension of classical ideas. The general public certainly sees the discovery of the structure and function of DNA as a confirmation of the classical view that the organism's characters are determined by information coded in the genes. Scientists and historians alike have presented the post-war developments in molecular genetics as a revolution that has unlocked the 'key of life'. Classical genetics introduced the basic concepts of hereditary determination, but molecular biology has shown us how the code is stored and how it is expressed. It is easy to fall into the trap of assuming that if we understand the genetic code, we understand the foundations of life itself and have the power to determine at will the characters of future living organisms. This in turn reinforces the social attitudes associated with classical genetics, including the view that a person's character is genetically predetermined. Determinism has also been extended by modern developments in evolution theory: some exponents of sociobiology argue that our social instincts have a genetic component which has been shaped by natural selection.

We shall see, however, that the popular image of the second revolution in the study of heredity is considerably oversimplified. The modern understanding of the genetic code has, in at least some respects, exposed weaknesses in the conceptual structure of classical genetics. The notion of a unit gene turns out to have no clear equivalent at the level of chromosomal DNA. The complexity of the process by which the information stored in the gene is transcribed into organic structures, coupled with the fact that much DNA has turned out to have no higher function, has made it extremely difficult to define a precise equivalent to the classical concept of the gene (Falk, 1986). More important still, exploration of the process by which the genetic information is expressed has opened up an area of study largely ignored by classical geneticists. It is now clear that the environment plays a significant role in determining how the genetic information shapes the growth of the organism. Some geneticists and many non-scientists are concerned to ensure that we do not lose sight of the fact that a human being is something more than the information stored in a genetic programme.

Molecular Biology and the Gene

The classical genetics of the 1920s and 30s treated the gene as a chromosomal unit somehow programmed to generate a particular character in the organism. There was only the vaguest notion of how the information might be coded in the gene, but most geneticists believed that it was stored in the complex structure of protein molecules. Confirmation that the true material of heredity was not protein but nucleic acid, specifically deoxyribonucleic acid or DNA, came only shortly before J. D. Watson and Francis Crick worked out the double-helix structure of the DNA molecule in 1953. There are several extensive accounts of the development of molecular genetics, including a highly personal statement by Watson himself (Watson, 1968; see also Olby, 1974; Portugal and Cohen, 1977; Judson, 1979).

The way in which the problem of the gene's physical structure was tackled imposed restrictions that at first seemed only to reinforce the classical geneticists' lack of interest in gene expression. In their attempt to understand the way in which genetic information is stored, biologists had to evade the complexities introduced by the process of development. Identification of the genetic role played by DNA was made possible by concentrating on ever-smaller organisms: fruit flies were replaced by bacteria and finally viruses. At this level, growth was irrelevant – since a virus is merely a naked gene – and biologists could concentrate on the transmission of the genetic material itself. But to understand how the genetic information is expressed, biologists had to move in exactly the opposite direction. They needed techniques developed not by classical geneticists, but by experimental embryologists who had retained an interest in the mechanics of the growth-process. French biologists – who had little or no background in classical genetics – were thus able to play a vital role in the emergence of molecular biology (Burian, Gayon and Zallen, 1988).

To sketch in an outline of how molecular genetics was created, we must return to the classical geneticists' concept of the nature and function of the gene. It has been argued that the geneticists of the 1930s had little real interest in this side of their theory, since they treated the material gene merely as a convenient symbol for analysing chromosomal behaviour (Stent, 1970). On this model, the emergence of molecular genetics marks a major break with the past. It is certainly true that many classical geneticists paid little attention to the problem of expression, and thus failed to ask how the gene might actually work. But most historians of genetics accept that this was an inevitable consequence of the limited techniques available for tackling the problem. Some classical geneticists *were* interested in the material nature of the gene, and proposed indirect methods of trying to study it. Their work suggests that the concept of a material gene was already central, and that a move towards studying its nature and function would begin as soon as appropriate techniques became available.

Under the microscope, chromosomes exhibit dark bands which can be related to the loci of individual genes ascertained by mapping. This phenomenon suggested that there were physical changes along the length of the chromosome that might be associated with the means of information-storage. As yet, however, the chemical nature of the genetic substance remained elusive. H. J. Muller and others hoped that an analysis of radiation-induced mutations would allow physicists to throw light on the molecular structure of the gene. Studying how the molecule could be changed by radiation might offer clues as to its chemical structure. Although the suggestion certainly indicates an interest in the structure of the gene, the method turned out to have severe limitations.

New possibilities were then opened up through the discovery of viruses. In 1917 F. Twort and F. d'Hérelle independently announced the discovery of bacteriophages (viruses which infect bacteria). At first there was considerable debate over the true nature of viruses, but by the 1930s geneticists such as Muller had begun to suspect that they were merely naked genes capable of duplication only when absorbed into a host cell. In 1936 W. M. Stanley was able to crystallize tobacco mosaic virus and study its chemical composition. He found that it consisted of 90 per cent protein and 10 per cent nucleic acid. On the basis of such figures, most biologists assumed that the active material of the virus, and hence presumably of genes, was the protein. The early work on viruses thus seemed to support what Olby (1974) calls the 'protein version of the central dogma' – the claim that protein molecules alone possess the degree of complexity necessary to act as a code specifying the complex structure of a biological organism.

The nucleic acids had been discovered in the cell nucleus during the 1870s, but their structure and function remained a mystery. A. Kossel began to identify the bases that form part of these large molecules, now known to be adenine, cytosene, guanine, plus uracil (in RNA) or thyamine (in DNA). Kossel also showed that carbohydrate was present, and in the 1920s Phoebus A. Levene identified this as ribose (hence the name ribonucleic acid, RNA) or deoxyribose (hence deoxyribonucleic acid, DNA). The situation was complicated at first because many biologists assumed that DNA occurred only in animals, RNA only in plants. Nor was there any way of knowing how the components of the molecule were arranged. Levene realized that the nucleic acid molecule was very large, but he did not believe that it was complex enough to store genetic information. His 'tetranucleotide hypothesis' pictured DNA as a long but essentially uniform polymer. This merely reinforced the belief that proteins were responsible for coding the genetic information.

At this point biolgists' attitudes toward DNA were influenced by studies of the apparently unrelated phenomenon of bacterial transformation. It was known that bacteria could be changed from an active to a passive form by the transfer of what was assumed to be genetic material. In the 1940s O. T. Avery's work on pneumonia bacteria showed that the agent of transformation was DNA. This raised the prospect that DNA and not protein was the active genetic material, although at first many biologists remained sceptical. Among the few to take the possibility seriously were the members of the 'phage group' centred on Max Delbrück, Alfred Hershey and Salvador Luria. They had begun work on bacteriophages (bacterial viruses) in the hope of understanding their genetic function. In 1952 Hershey and Martha Chase used radioactive labelling to show that the active constituent of the virus is DNA, the protein forming only a protective coat that facilitates entry of the genetic material into the host cell. Erwin Chargaff now suggested that DNA was not a regular polymer, and that the arrangement of the bases in its molecular structure might provide a means of encoding genetic information. At this point a young American member of the phage group, James D. Watson, travelled to Europe in the hope of finding physical evidence that would reveal the exact structure of DNA.

There were others working on DNA by now, including Linus Pauling, who used his knowledge of atomic physics to explain the structure of the protein molecule. One vital source of information on the structure of large molecules was their X-ray diffraction patterns. This technique had been pioneered in Britain by Sir Lawrence Bragg, and it was to make use of this kind of information that Watson eventually moved to Cambridge. Here he teamed up with Francis Crick, and the pair soon became interested in constructing a model for the structure of DNA. They made use of X-ray diffraction patterns obtained by Maurice Wilkins, who was working in uneasy collaboration with Rosalind Franklin in London. Wilkins believed that his diffraction patterns could best be explained by a spiral structure, but it was only after several false starts that Watson and Crick finally put together the concept of the double helix. In their classic paper of 1953 (reprinted in Judson, 1979, pp. 196–8) they described a structure in which the carbohydrate chains serve as an external skeleton bonded together by the bases. The exact arrangement of the bases would represent the genetic code. Their paper noted briefly that the model would provide a 'copying mechanism for the genetic material'. This followed from the fact that the bases could only link up across the helix in a certain way, adenine bonding to thyamine and guanine to cytosine. If the two chains unrolled, the bases would pair up in the appropriate way to recreate the other side of the helix, thus generating two molecules from one. The exact replication of the genetic information from cell to cell could thus be explained.

The double helix model showed how the genetic information could be encoded in the exact sequence of base-pairs, and how that information was transmitted from cell to cell by chromosome duplication. At one level, it could be argued that molecular biology has merely realized the potential inherent within classical genetics. It has given us a better understanding of the nature of the genetic units postulated by the geneticists of the 1930s. And yet there is a sense in which the concepts of that earlier period have now been shown to be unworkable. The new discoveries have revealed processes so complex that it is scarcely possible to ask what a gene (in the classical sense of the term) really is. At the simple level, a gene is a section of DNA coded to produce a particular protein at some point in the developmental process. But we have now discovered that much DNA is repetitive, while some serves no purpose in individual development ('junk DNA'). The same section of DNA can serve both a structural and a regulatory purpose, and even in a segment comprising a structural gene there is material that must be deleted before translation can take place. In the end, it is extremely difficult to come up with a definition that satisfies modern requirements and at the same time corresponds unambiguously to the classical notion of the unit gene. Molecular genetics has expanded the concepts of the classical phase to an extent that has made them unrecognizable, although the term 'gene' still figures in most nonspecialist vocabularies.

Despite these conceptual problems, the new techniques have immense practical applications. Molecular genetics has allowed us to understand some of the most basic processes of life, and has opened up the prospect that we may actually take control of those processes. Better knowledge of how the genetic code actually functions has created a new biological technology. It has become possible to identify the genes responsible for particular characters and to transfer genetic material from one organism to another. Recombinant DNA research has laid the foundations for a new industrial revolution, based on our ability to manufacture micro-organisms designed to perform useful chemical reactions. Small wonder that some critics accuse the geneticists of playing God by creating new forms of life. There is also grave concern over the possible dangers that might arise from the release of artificially produced micro-organisms into the natural environmnent. Strict limitations are imposed on recombinant DNA work in an attempt to head off potential criticisms from this source.

Perhaps even more thought-provoking are the medical impli-cations of the new genetics. Thanks to better knowledge of the human genome, it is now possible to identify the position of the genes responsible for the bodily malfunctions at the heart of some hereditary diseases. A project is under way to sequence the whole human genome, providing a complete representation of the genetic information from which a human being is constructed. The techniques now becoming available will allow the insertion of new DNA to 'correct' the malfunctions in somatic cells, thus effectively negating the harmful aspects of a person's genetic inheritance. No one is as yet discussing the possibility of altering the DNA in human germ cells, thus permanently altering the gene pool of the human race. Nevertheless, if the appropriate techniques become available, a debate may open up on the advisability of simply eliminating harmful genes from the human race altogether. Such a project would have wide implications raising the gravest moral issues. Geneticists are well aware of the bad name their discipline gained through its association with the old eugenics movement. Future biotechnology may allow eugenic ideals to be realized without restricting the 'unfit' from breeding altogether. Yet the moral problem would remain unchanged: who decides which gene is good and which is bad? There is a slippery slope leading from the correction of obvious genetic defects down to the nightmare of a human race that has been artificially designed to meet the re-quirements of a particular ideology.

The Gene and the Organism

The debates over the potentials and dangers of recombinant DNA work have to some extent obscured the fact that there is another side to modern genetics dealing with the mechanism by which the gene's information is expressed in the developing organism. By focusing on the DNA itself, we uphold the traditional determinist view in which the genetic information is regarded as sufficient to specify the whole character of the organism. The moral dilemmas presented by the new genetics can to some extent be highlighted by pointing to the artificiality of this determinist perspective. The organism – particularly the human organism – is something more than a physical manifestation of genetic information, and one way of emphasizing this fact is to look at the complexity of the process by which the genetic information is expressed.

The double helix model showed how the genetic information could be encoded in the exact sequence of the base-pairs. It did not, however, explain how the information thus stored was decoded to control the formation of the proteins which make up the living body. From the start, Francis Crick suspected that the information would be translated into the construction of protein molecules using RNA as an intermediate. He thus pioneered the 'central dogma' of molecular biology: the information-flow between genes and organisms is strictly a one-way affair, from DNA to RNA to proteins. By the 1960s the central dogma was firmly established. Biologists were convinced that the DNA molecule serves as a template for the construction of RNA which in turn acts as a template for the construction of the proteins that govern the structure and functioning of the organism. There was no obvious way in which proteins could influence the structure of the DNA to provide a reverse flow or feedback from the organism to its genetic material. The central dogma thus reinforces the philosophy of preformationism established by classical genetics. It was all too easy to assume that the whole character of the future organism was somehow encapsulated within the fertilized ovum.

By virtue of the central dogma, molecular genetics has sustained the hostile attitude adopted by Weismann and the classical geneticists towards Lamarckism. The organism is built using information supplied by its genetic inheritance, but changes acquired by the adult organism cannot be fed back into the genes it transmits to future generations. Genetic mutation is the only natural source of new characters. Efforts are still occasionally made to postulate a flow of information in the reverse direction, thus readmitting Lamarckism and allowing the organisms' behaviour to influence the future of their species. Lamarckism is too attractive a prospect to be banished from people's minds by a simple demonstration of how the genetic code actually works. Yet it is significant that the most recent effort in this direction, the 'somatic selection' hypothesis of Ted Steele (1979), evaded rather than violated the central dogma. Steele argued that the immune system works by selecting useful mutations within the somatic cells, and suggested that viruses might sometimes be able to transmit this somatic DNA into the germ cells. His work was bitterly criticized at the time by biologists who doubted his experimental results and rejected his hypothetical mechanism as implausible.

The central dogma thus remains essentially intact, and its assumption that the information flows in only one direction from the genes to the organism helps to create the popular impression that the genes completely determine the organism's character. We often think of the organism as having a genetic 'blueprint' specifying all the information necessary to construct it. Yet even within the boundaries of the central dogma, it can be shown that the assumption of rigid determinism is a vast oversimplification. The old idea that each gene uniquely and absolutely determines a single character is quite unrealistic. Even without the possibility of a feedback from the organism to its genes, the process of gene-expression is so complex that the organism must be seen as something more than the physical manifestation of a collection of genetic units. Not only do the genes interact with one another in the course of the growth process, but the environment within which the growth takes place can also exert considerable influence over the end product. The process of manufacturing an organism – like the process of converting a blueprint into a finished product – requires more than the information expressed in the blueprint itself.

To uncover the details of gene expression, modern biologists have had to make use of techniques quite unlike those of classical genetics. In the post-war years, the French biologist Boris Ephrussi had already begun to tackle this problem by adopting an embryo-logist's approach, working backwards from the end-products of the growth process. Along with George Beadle and Edward Tatum,

Ephrussi explored the possibility that the genes work by controlling the production of the enzymes responsible for protein formation. Bacterial studies confirmed that RNA acted as an intermediate template, formed by the DNA of the cell nucleus to control the synthesis of proteins in the ribosomes. In 1969 Francois Jacob and Jacques Monod introduced the concept of 'messenger RNA' to denote the function of RNA as a transient intermediate between the genes and the growing organism. To explain how the formation of proteins is co-ordinated, they proposed that sections of the genome have their activity switched on and off as units by the action of 'regulator genes'.

French biologists such as Ephrussi, Jacob and Monod were able to play an important role in uncovering the mechanism of gene expression because they came from a background in which the question of growth had not been dismissed as irrelevant. For molecular biologists to create our modern understanding of the genetic code and its means of expression, they had to absorb studies of growth based on techniques pioneered by scientists who had not been influenced by the tradition of classical genetics (Burian, Gayon and Zallen, 1988). But those outside the classical tradition did not share the view that each gene absolutely determines a particular character of the organism. European scientists in particular have tried to keep alive the older developmental tradition in which growth is something more than the mechanical expression of pre-existing information. In some cases they have openly challenged the determinist views implied by English-speaking geneticists, and certainly their work suggests another dimension that must be taken into account in discussing the implications of modern biology.

In some respects, the limitations of simple determinism are exposed more obviously through the interaction between genetics and evolution theory. By its very nature, evolutionism must address itself to the nature of the organism, which has to survive and reproduce within a real-world environment. Yet population genetics was introduced in part as a means of reducing evolution to changes in gene frequency brought about by the action of selection. Evolutionists have thus been forced into debates over the units of selection: is selection acting on populations, on the phenotypes of individual organisms, on gene-interaction systems, or on the genes themselves? Some of the original formulations of the genetical theory of natural selection were based on 'beanbag genetics', which assumed that new characters were added to the gene pool by individual mutations and were evaluated as units by natural selection. More recently, Richard Dawkins' concept of the 'selfish gene' has been widely (if erroneously) interpreted as implying that the organism is nothing more than a puppet in the hands of genes determined to win the game of evolution by dominating the species.

Yet from the start there were some biologists who pointed out that this approach ignored the complex process of development intervening between the genes and the production of the mature organism. The work of Sewall Wright was based on the assumption that selection must act on gene-interaction systems built up within local populations. On this model, no gene can be evaluated as a unit because its effect depends upon all the possible interaction-systems that may arise when it is brought into combination with other genes already in the population. The individual organism must, then, be seen as something more than a mosaic of unit characters, and one cannot always predict the effect that a particular gene will have.

Far from advocating a beanbag view of genetics, Richard Dawkins (1986) has pointed out that our modern knowledge of the way in which genes are expressed has rendered the concept of a genetic blueprint obsolete. He suggests that it would be better to visualize the genetic information as a *recipe* or series of instructions for creating an organism. In a blueprint there is a one-to-one correspondence between each element of the plan and its manifestation in the real world. But genes do not really encode for characters: they produce sequences of proteins which, interacting under suitable conditions, will result in the construction of a new organism. As Dawkins points out, it is precisely because the genes form a recipe and not a blueprint that Lamarckism is impossible. There is no conceivable mechanism that could translate a change in bodily structure into an equivalent change in the genetic recipe. Contrary to what was expected by biologists at the turn of the century, preformation via a genetic blueprint would make Lamarckism more plausible by specifying which part of the genome would have to be changed to mimic an environmental effect. An epigenetic process of growth based on a genetic recipe will not allow modifications resulting from changed conditions to affect the recipe itself. Selection alone can pick out mutations which, in one combination or another, offer a better recipe in a changed environment.

As a Darwinist, Dawkins is interested in those cases where the effects of the genetic recipe remain quite predictable. He shares the traditional Darwinian view that only those mutations which have a very small effect on the organism are likely to have beneficial consequences. Yet there has always been a small counter-movement within biology which suspected that the population geneticists had even more seriously underestimated the role of the developmental process. We have seen (Chapter 7) how Richard Goldschmidt campaigned against Darwinism during the 1930s and 40s, arguing that micromutations could never do anything more than produce minor changes within the species. Goldschmidt appealed to larger, saltative changes to explain the origin of new characters. Orthodox Darwinians ridiculed Goldschmidt's concept of the 'hopeful monster', yet the idea is receiving new support among biologists even today (Ho and Saunders, 1984). The theory of 'epigenetic evolution' is based on the assumption that in some cases a quite small mutation may have major consequences because at some crucial point it switches the developmental process onto an entirely new track. The growth process itself may be able to co-ordinate the effects of the new gene so that the end result is viable – although significantly different to the parent form.

Epigenetic evolutionism is an extreme manifestation of the view that the genes do not determine the structure of the organism in an absolutely predictable way. By postulating a network of potentially viable developmental pathways, with crucial points of divergence where a slight change may have an ever-widening series of consequences, it reintroduces the developmental tradition that genetics was once thought to have vanquished. The concept of a genetic recipe has less extreme consequences, but nevertheless reminds us that the genes are not always the determinants of single characters. Everyone knows that the taste of a cake depends not just on the recipe, but on the skill of the cook, the quality of the ingredients and the temperature of the oven. In other words, the environment does play a role in shaping the growth of the organism. The organism must be something more than a physical expression of the information encoded in the DNA it inherits, and this is a point with major consequences for our evaluation of modern biology's social implications.

Consider first the medical applications of recombinant DNA research. It is perfectly obvious that many illnesses have a genetic component, and simple determinism would encourage us to tackle the problem by eliminating the genes that are responsible. But in some cases the disease may result from a genetic predisposition which is only manifested in a certain environment. Should we not then be tackling the problem of identifying the environmental rather than the genetic factors involved? Even when a single gene is known to cause a disease, it must be remembered that genes seldom have a single effect on the growing organism. Sickle-cell anaemia is widespread in the tropics because the gene responsible also confers immunity to malaria. Biologists will have to be very careful that they do not cut out a 'harmful' gene which actually has another, beneficial effect.

The role of the environment is particularly important when it comes to assessing the inheritance of mental capacities. There is growing evidence that some mental illnesses have a genetic component, but we must beware of the temptation to identify the carriers of certain genes as inherently inferior. The claim that an individual's level of intelligence is genetically determined is still advanced by a few educationalists, yet many psychologists and social workers – along with some geneticists – disagree (Rose, Kamin and Lewontin, 1984). The individual is a product of both nature *and* nurture, of both its genetic inheritance and the environment in which that information is expressed. Even if it were possible to define certain persons as 'inferior' (e.g. because of below-average intelligence), they could not be labelled as *genetically* inferior because their environment and upbringing will have played a major role in shaping their character. There are, in any case, major doubts as to whether IQ testing can really distinguish between native intelligence and the effects of education. To the extent that the concept of a genetic blueprint is perceived as implying determinism of human character, it needs to be abandoned in the light of modern ideas on how the genetic potential is unfolded.

The same point applies in the case of human behaviour. In the last few decades the science of 'sociobiology' has had great success in explaining the apparently altruistic instincts of some animals in terms of Darwinian natural selection (Caplan, 1978). Using the concept of the 'selfish gene', sociobiologists have shown that individuals can be programmed with instincts that threaten their own survival, if their altruistic behaviour helps their genetic relatives to breed. The founder of sociobiology, E. O. Wilson, has expressed the view that at least some aspects of human behaviour might turn out to have a similar genetic component, programmed into us by evolution. By exploiting the implications of evolution theory, sociobiology extends genetic determinism to include the instincts which, it is claimed, govern our behaviour and thus determine the structure of our society.

Social scientists are bitterly hostile to this renewed effort by biologists to extend their influence into the area of the human sciences. They advocate a purely environmentalist view in which the human mind is a blank slate upon which culture will impose the values of the society in which the individual is raised. This may be going too far, but the social scientists are right to point out that heredity endows us with a brain that is capable of learning a wide range of skills and customs. We should be very careful of the tendency to assume that all mental functions and faculties are genetically predetermined. Such claims all too easily play into the hands of those who wish to perpetuate social inequalities by arguing that they are 'natural'. Modern genetics offers immense opportunities for improving human life, but history shows that efforts to stress the role of heredity in human affairs have all too often been based upon ideological rather than scientific considerations. We need to think very carefully about how we make use of the new tools, and the new ideas, that the biologists have created. Despite all the claims that the structure of DNA represents the 'secret of life', the public must be told that biology does not endorse the hereditarian attitudes promoted in the heyday of classical genetics.

Bibliography

ADAMS, Mark B. (1968) 'The Foundations of Population Genetics: Contributions of the Chetverikov School', *J. Hist. Biology*, 1: 23–39.

ADELMANN, Howard B. (1966) *Marcello Malpighi and the Evolution of Embryology*, 5 vols. Ithaca, NY: Cornell University Press.

ALLEN, Garland E. (1968) 'Thomas Hunt Morgan and the Problem of Natural Selection', *J. Hist. Biology*, 1: 113–39.

—— (1969) 'Hugo De Vries and the Reception of the Mutation Theory', *J. Hist. Biology*, 2: 55–87.

—— (1974) 'Opposition to the Mendelian-Chromosome Theory: the Physiological and Developmental Genetics of Richard Goldschmidt', *J. Hist. Biology*, 7: 49–92.

—— (1975) *Life Science in the Twentieth Century*. New York: Wiley.

—— (1976) 'Genetics, Eugenics and Society', *Social Studies of Science*, 6: 105–22.

—— (1978) *Thomas Hunt Morgan: the Man and his Science*. Princeton University Press.

—— (1981) 'Morphology and Twentieth-Century Biology: a Response', *J. Hist. Biology*, 14: 159–76.

—— (1986a) 'T. H. Morgan and the Split between Embryology and Genetics, 1910–35', in Horder, Witkowski and Wylie (1986) pp. 113–46.

—— (1986b) 'The Eugenics Records Office at Cold Spring Harbor, 1910–1940', *Osiris*, n.s. 2: 225–64.

ANDERSON, Lorin (1982) *Charles Bonnet and the Order of the Known*. Dordrecht: D. Reidel.

BAER, Karl Ernst von (1828) *Über Entwickelungsgeschichte der Thiere, Beobachtung und Reflexion: Erster Teil*. Königsberg.

BAJEMA, Carl Jay (1977) *Eugenics, Then and Now*. Stroudsberg, Pa.: Dowden, Hutchinson and Ross.

BANNISTER, Robert C. (1979) *Social Darwinism: Science and Myth in Anglo-American Social Thought*. Philadelphia: Temple University Press.

BARNES, Barry, and EDGE, David (eds) (1982) *Science in Context: Readings in the Sociology of Science*. Milton Keynes: Open University Press.

BARNES, Barry, and SHAPIN, Steven (eds) (1979) *Natural Order: Historical Studies of Scientific Culture*. Beverly Hills and London: Sage Publications.

BATESON, Beatrice (1928) *William Bateson, F.R.S., Naturalist*. Cambridge University Press.

BATESON, William (1894) *Materials for the Study of Variation, Treated with Especial Regard to Discontinuity in the Origin of Species*. London: Macmillan.

—— (1902) *Mendel's Principles of Heredity: a Defence*. Cambridge University Press.

—— (1909) *Mendel's Principles of Heredity*. Cambridge University Press.

—— (1913) *Problems of Genetics*. Reprinted New Haven: Yale University Press, 1979.

—— (1914) 'President's Address', *Report of the British Association for the Advancement of Science*, 1914 meeting: 3–38.

—— (1928) *The Scientific Papers of William Bateson*, ed. R. C. Punnett, 2 vols. Cambridge University Press.

BAXTER, Alice, and FARLEY, John (1979) 'Mendel and Meiosis', *J. Hist. Biology*, 12: 137–73.

BENNETT, J. H. (ed.) (1983) *Natural Selection, Heredity and Eugenics*. Oxford University Press.

BENSON, Keith R. (1981) 'Problems of Individual Development: Descriptive Embryological Morphology in America at the Turn of the Century', *J. Hist. Biology*, 14: 115–28.

BLIXT, Stig (1975) 'Why Didn't Gregor Mendel Find Linkage?', *Nature*, 256: 206.

BLOOR, David (1976) *Knowledge and Social Imagery*. London: Routledge and Kegan Paul.

BOESIGER, Ernest (1980) 'Evolutionary Biology in France at the Time of the Evolutionary Synthesis', in Mayr and Provine (1980), pp. 309–20.

BONNET, Charles (1779) *Oeuvres d'histoire naturelle et de philosophie*, 19 vols. Neuchatel.

BOVERI, Theodor (1904) *Ergebnisse über die Konstruction der chromatischen Substanz des Zellkernes*. Jena: Fischer.

BOWLER, Peter J. (1971) 'Preformation and Pre-existence in the Seventeenth Century: a Brief Analysis', *J. Hist. Biology*, 4: 221–44.

—— (1973) 'Bonnet and Buffon: Theories of Generation and the Problem of Species', *J. Hist. Biology*, 6: 259–81.

—— (1974) 'Darwin's Concepts of Variation', *J. Hist. Medicine*, 29: 196–212.

—— (1978) 'Hugo De Vries and Thomas Hunt Morgan: the Mutation Theory and the Spirit of Darwinism', *Annals of Sci.*, 35: 55–73.

—— (1983) *The Eclipse of Darwinism: Anti-Darwinian Evolution Theories in the Decades around 1900*. Baltimore: Johns Hopkins University Press.

—— (1984a) *Evolution: the History of an Idea*. Berkeley and Los Angeles: University of California Press.

—— (1984b) 'E. W. MacBride's Lamarckian Eugenics and its Implications for the Social Construction of Scientific Knowledge', *Annals of Sci.*, 41: 245–60.

—— (1986) *Theories of Human Evolution: a Century of Debate, 1844–1944*. Baltimore: Johns Hopkins University Press and Oxford: Basil Blackwell.

—— (1988) *The Non-Darwinian Revolution: Reinterpreting a Historical Myth*. Baltimore: Johns Hopkins University Press.

BOX, Joan Fisher (1978) *R. A. Fisher: the Life of a Scientist*. New York: Wiley.

BRANNIGAN, Augustine (1979) 'The Reification of Mendel', *Social Studies of Sci.*, 9: 423–54.

BRUSH, Stephen G. (1978) 'Nettie M. Stevens and the Discovery of Sex Determination by Chromosomes', *Isis*, 69: 163–72.

BUCKLE, Henry Thomas (1857) *History of Civilization in England*, vol. 1. London.

BUFFON, Georges Louis Leclerc, Comte de (1749–67) *Histoire naturelle, générale et particulière*, 15 vols. Paris.

—— (1774–89) *Histoire naturelle. Supplément*, 7 vols. Paris.

—— (1785) *Natural History*, trans. William Smellie, 2nd edn, 7 vols. London.

BUICAN, Denis (1984) *Histoire de la génétique et d'évolutionisme en France*. Paris: PUF.

BURIAN, Richard M. (in press) 'French Contributions to the Research Tools of Molecular Genetics, 1945–1960', *Revue de synthèse*.

BURIAN, R. M., GAYON, J. and ZALLEN, D. (1988) 'The Singular Fate of Genetics in the History of French Biology', *J. Hist. Biology*, 21: 357–402.

BURKHARDT, Richard W., Jr (1977) *The Spirit of System: Lamarck and Evolutionary Biology*. Cambridge, Mass.: Harvard University Press.

BUTLER, Samuel (1920) *Unconscious Memory*, 3rd edn. London: Fifield.

CALLENDER L. A. (1988) 'Gregor Mendel – an Opponent of Descent with Modification', *History of Science*, 26: 41–75.

CAMPBELL, M. (1980) 'Did De Vries Discover the Law of Segregation Independently?', *Annals of Sci.*, 37: 639–55.

CAPLAN, Arthur L. (ed.) (1978) *The Sociobiology Debate*. New York: Harper and Row.

CARLSON, Elof A. (1966) *The Gene: a Critical History*. Philadelphia: Saunders.

—— (1974) 'The *Drosophila* Group: the Transition from the Mendelian Unit to the Individual Gene', *J. Hist. Biology*, 7: 31–48.

—— (1981) *Genes. Radiation and Society: the Life and Work of H. J. Muller*. Ithaca, NY: Cornell University Press.

CASSIRER, Ernst (1951) *The Philosophy of the Enlightenment*. Princeton University Press.

CASTLE, W. E. (1911) *Heredity in Relation to Evolution and Animal Breeding*. New York: Appleton.

CHAMBERS, Robert (1844) *Vestiges of the Natural History of Creation*. London: Churchill.

CHURCHILL, Frederick B. (1968) 'August Weismann and a Break from Tradition', *J. Hist. Biology*, 1: 91–112.

—— (1970) 'Hertwig, Weismann and the Meaning of Reduction Division', *Isis*, 61: 429–58.

—— (1973) 'Chabry, Roux and the Experimental Method in Nineteenth-Century Embryology', in Ronald N. Giere and Richard S. Westfall (eds), *Foundations of Scientific Method: the Nineteenth Century*, Bloomington: Indiana University Press, pp. 161–205.

—— (1974) 'William Johanssen and the Genotype Concept', *J. Hist. Biology*, 7: 5–30.

—— (1986) 'Weismann, Hydromedusae and the Biogenetic Imperative, a Reconsideration', in Horder, Witkowski and Wylie (1986) pp. 7–33.

—— (1987) 'From Heredity Theory to *Verebung*: the Transmission Problem, 1850–1915', *Isis*, 78: 337–64.

CLARK, Ronald W. (1969) *JBS: the Life and Work of J. B. S. Haldane*. New York: Coward-McCann.

COCK, A. G. (1973) 'William Bateson, Mendelism and Biometry', *J. Hist. Biology*, 6: 1–36.

COLE, F. J. (1930) *Early Theories of Sexual Generation*. Oxford: Clarendon Press.

COLEMAN, William (1965) 'Cell, Nucleus and Inheritance: an Historical Study', *Proc. Am. Phil. Soc.*, 109: 124–58.

—— (1970) 'Bateson and Chromosomes: Conservative Thought in Science', *Centaurus*, 15: 228–315.

CORRENS, Carl (1900) 'G. Mendel's Regel über das Verhalten der Nachkommenschaft der Rassenbastarde', *Berichte der deutschen botanischen Gesselschaft*, 18: 158–68.

COWAN, Ruth Schwartz (1972a) 'Francis Galton's Contributions to Genetics', *J. Hist. Biology*, 5: 389–412.

—— (1972b) 'Francis Galton's Statistical Ideas: the Influence of Eugenics', *Isis*, 63: 509–28.

—— (1977) 'Nature and Nurture: the Interplay of Biology and Politics in the Work of Francis Galton', *Stud. Hist. Biology*, 1: 133–208.

CRAVENS, Hamilton (1978) *The Triumph of Evolution: American Scientists and the Heredity-Environment Controversy, 1900–1941*. Philadelphia: University of Pennsylvania Press.

DARDEN, Lindley (1976) 'Reasoning in Scientific Change: Charles Darwin, Hugo De Vries and the Discovery of Segregation', *Stud. Hist. and Phil. Sci.*, 7: 89–126.

—— (1977) 'William Bateson and the Promise of Mendelism', *J. Hist. Biology*, 10: 87–106.

DARWIN, Charles (1859) *On the Origin of Species by Means of Natural Selection*. London: John Murray.

—— (1868) *The Variation of Animals and Plants under Domestication*, 2 vols. London: John Murray.

—— (1887) *The Life and Letters of Charles Darwin*, ed. Francis Darwin, 3 vols. London: John Murray.

—— (1987) *Charles Darwin's Notebooks, 1836–1844*, ed. Paul H. Barrett *et al.* Cambridge University Press.

DARWIN, Erasmus (1794–6) *Zoonomia, or the Laws of Organic Life*, 2 vols. London.

DAWKINS, Richard (1986) *The Blind Watchmaker*. Harlow: Longmans Scientific and Technical.

DE MARRAIS, Robert (1974) 'The Double-edged Effect of Francis Galton: a Search for the Motives in the Biometrician–Mendelian Debate', *J. Hist. Biology*, 7: 141–74.

DESMOND, Adrian (1982) *Archetypes and Ancestors: Palaeontology in Victorian London, 1850–1875*. London: Blond and Briggs.

—— (1984) 'Robert E. Grant: the Social Predicament of a Pre-Darwinian Evolutionist', *J. Hist. Biology*, 17: 189–223.

DE VRIES, Hugo (1900a) 'Sur la loi de disjonction des hybrides', *Comptes Rendus, Acad. Sci. Paris*, 130: 845–7.

—— (1900b) 'Das Spaltungsgesetz der Bastarde', *Berichte der deutschen botanischen Gesselschaft*, 18: 83–90.

—— (1910a) *Intracellular Pangenesis*, trans. C. Stuart Gager. Chicago: Open Court.

—— (1910b) *The Mutation Theory: Experiments and Observations on the Origin of Species in the Vegetable Kingdom*, trans. J. B. Farmer and A. D. Darbyshire, 2 vols. London: Kegan Paul.

DI GREGORIO, Mario (1984) *T. H. Huxley's Place in Natural Science*. New Haven: Yale University Press.

DIJKSTERHUIS, E. J. (1961) *The Mechanization of the World Picture*. Oxford: Clarendon Press.

DOBZHANSKY, Theodosius (1937) *Genetics and the Origin of Species*. New York: Columbia University Press.

DUNN, L. C. (ed.) (1951) *Genetics in the Twentieth Century*. New York: Macmillan.

—— (1965) *A Short History of Genetics*. New York: McGraw-Hill.

EISELEY, Loren (1958) *Darwin's Century: Evolution and the Men Who Discovered It.* New York: Doubleday.

EVANS, Brian, and WAITES, Bernard (1981) *IQ and Mental Testing: an Unnatural Science and its Social History.* London: Macmillan.

FALK, Raphael (1986) 'What is a Gene?', *Stud. Hist. and Phil. of Sci.*, 17: 133–73.

FANCHER, R. (1983) 'Francis Galton's African Ethnology and its Role in the Development of his Ideas', *Brit. J. Hist. Sci.*, 16: 67–79.

FARBER, Paul L. (1972) 'Buffon and the Problem of Species', *J. Hist. Biology*, 5: 259–84.

FARLEY, John (1982) *Gametes and Spores: Ideas about Sexual Reproduction, 1750–1914.* Baltimore: Johns Hopkins University Press.

FARRALL, Lyndsay A. (1979) 'The History of Eugenics: a Bibliographical Review', *Annals of Sci.*, 36: 111–23.

FISHER, Ronald Aylmer (1918) 'The Correlation between Relatives on the Supposition of Mendelian Inheritance', *Trans. Roy. Soc. Edinburgh*, 52: 399–433.

—— (1930) *The Genetical Theory of Natural Selection.* Oxford: Clarendon Press.

—— (1936) 'Has Mendel's Work been Rediscovered?', *Annals of Sci.*, 1: 115–37.

FORREST, D. (1974) *Francis Galton: the Life and Work of a Victorian Genius.* New York: Tapplinger.

FROGGATT, P and NEVIN, N. C. (1971) 'The "Law of Ancestral Heredity" and the Mendelian-Ancestrian Controversy in England, 1889–1900', *J. Med. Genetics*, 8: 1–36.

GALTON, Francis (1865) 'Hereditary Talent and Character', *Macmillan's Magazine*, 12: 157–66 and 318–27.

—— (1869) *Hereditary Genius: an Inquiry into its Laws and Consequences.* London: Macmillan.

—— (1871) 'Experiments on Pangenesis', *Proc. Roy. Soc. Lond.*, 19: 393–410.

—— (1883) *Inquiries into Human Faculty and Development.* New York: Macmillan.

—— (1889) *Natural Inheritance.* London: Macmillan.

—— (1908) *Memories of my Life.* London: Methuen.

GASKING, Elizabeth (1967) *Investigations into Generation, 1651–1828.* London: Hutchinson.

GAY, Peter (1966–9) *The Enlightenment: an Interpretation*, 2 vols. New York: Alfred A. Knopf.

GEISON, Gerald (1969) 'Darwin and Heredity: The Evolution of his Hypothesis of Pangenesis', *J. Hist. Medicine*, 24: 375–411.

GILBERT, Scott F. (1978) 'The Embryological Origins of the Gene Theory', *J. Hist. Biology*, 11: 307–51.

—— (1988) 'Cellular Politics: Ernest Everett Just, Richard B. Goldschmidt, and the Attempt to Reconcile Embryology and Genetics', in R. Rainger, K. R. Benson and J. Maienschein (eds), *The American Development of Biology*. Philadelphia: University of Pennsylvania Press, pp. 311–46.

GILLISPIE, Charles C. (1951) *Genesis and Geology*. Reprinted New York: Harper, 1959.

—— (1959) 'Lamarck and Darwin in the History of Science', in Glass *et al.* (eds) (1959) pp. 265–91.

GLASS, Bentley (1959a) 'Maupertuis, Pioneer of Genetics', in Glass *et al.* (eds) (1959) pp. 51–83.

—— (1959b) 'Heredity and Variation in the Eighteenth-Century Concept of the Species', in Glass *et al.* (eds) (1959) pp. 144–72.

GLASS, Bentley, *et al.* (eds) (1959) *Forerunners of Darwin, 1745–1859*. Baltimore: Johns Hopkins University Press.

GOLDSCHMIDT, Richard (1940) *The Material Basis of Evolution*. Reprint (1982) introd. S. J. Gould. New Haven: Yale University Press.

GOULD, Stephen J. (1977) *Ontogeny and Phylogeny*. Cambridge, Mass.: Harvard University Press.

—— (1981) *The Mismeasure of Man*. New York: Norton.

HAECKEL, Ernst. (1876a) *The History of Creation*, 2 vols. New York: Appleton.

—— (1876b) *Die Perigenesis der Plastidule*. Berlin: Reimer.

—— (1879) *The Evolution of Man*, 2 vols. New York: Appleton.

HALDANE, J. B. S. (1932) *The Causes of Evolution*. London: Longmans.

—— (1938) *Heredity and Politics*. London: Allen and Unwin.

HALLER, John S. (1975) *Outcasts from Evolution: Scientific Attitudes of Racial Inferiority, 1859–1900*. Urbana: University of Illinois Press.

HALLER, Mark H. (1963) *Eugenics: Hereditarian Attitudes in American Thought*. New Brunswick, NJ: Rutgers University Press.

HAMBURGER, Viktor (1988) *The Heritage of Experimental Embryology: Hans Spemann and the Organizer*. Oxford University Press.

HAMPSON, Norman (1968) *The Enlightenment*. Harmondsworth: Penguin Books.

HARAWAY, Donna (1976) *Crystals, Fabrics and Fields: Metaphors of Organicism in Twentieth-Century Biology*. New Haven: Yale University Press.

HARVEY, William (1965) *The Works of William Harvey*. New York: Johnson Reprint Corporation.

HARWOOD, Jonathan (1984) 'The Reception of Morgan's Chromosome Theory in Germany', *Medizin historisches Journal*, 19: 3–32.

—— (1985) 'Genetics and the Evolutionary Synthesis in Interwar Germany', *Annals of Sci.*, 42: 279–301.

—— 'National Styles in Science: Genetics in Germany and the United States between the World Wars', *Isis*, 78: 390–414.

HEARNSHAW, L. S. (1979) *Cyril Burt, Psychologist*. Ithaca, NY: Cornell University Press.

HEMPEL, Carl (1966) *Philosophy of Natural Science*. Englewood Cliffs, NJ: Prentice Hall.

HENFREY, A. and HUXLEY, T. H. (eds) (1853) *Scientific Memoirs Selected from the Transactions of Foreign Academies* . . ., London, reprinted New York: Johnson Reprint Corporation, 1966.

HERTWIG, Oscar (1896) *The Biological Problem of Today: Preformation or Epigenesis?* London: Heinemann.

HO, Mae-Wan, and SAUNDERS, P. T. (1984) *Beyond Neo-Darwinism: an Introduction to the New Evolutionary Paradigm*. London: Academic Press.

HODGE, M J. S. (1971) 'Lamarck's Science of Living Bodies', *Brit. J. Hist. Sci.*, 5: 323–52.

—— (1972) 'The Universal Gestation of Nature: Chambers' *Vestiges* and *Explanations*', *J. Hist. Biology*, 5: 127–52.

—— (1985) 'Darwin as a Lifelong Generation Theorist', in Kohn (1985): 204–44.

HOLBACH, Paul Heinrich Dietrich, Baron d', (1821) *Système de la Nature*. Reprinted Hildesheim: Georg Olms, 1966, 2 vols.

HOFSTADTER, Richard (1955) *Social Darwinism in American Thought*, revised edn. New York: George Brazillier.

HORDER, T. J. and WEINDLING, P. (1985) 'Hans Spemann and the Organizer', in Horder, Witkowski and Wylie (1986) pp. 183–242.

HORDER, T. J., WITKOWSKI, J. A. and WYLIE, C. C. (eds) (1986) *A History of Embryology*. Cambridge University Press.

HULL, David L. (ed.) (1973) *Darwin and his Critics*. Cambridge, Mass.: Harvard University Press.

—— (1978) 'Sociobiology: a Scientific Bandwagon or a Travelling Medicine Show?' in M. S. Gregory *et al.* (eds) *Sociobiology and Human Nature*. San Francisco: Jossey-Bass, pp. 136–63.

HUXLEY, Julian (1942) *Evolution: the Modern Synthesis*. London: Allen and Unwin.

ILTIS, Hugo (1932) *Life of Mendel*. Reprinted New York: Hafner, 1966.

JENKIN, Fleeming (1867) 'The Origin of Species', *North British Review*, 46: 277–318.

JENSEN, Arthur S. (1969) 'How Much can we Boost IQ and Scholastic Achievement?', *Harvard Educational Review*, 33: 1–123.

—— (1972) *Genetics and Education*. New York: Harper and Row.

JOHANNSEN, Wilhelm (1911) 'The Genotype Conception of Heredity', *American Naturalist*, 45: 129–59.

—— (1955) 'Concerning Heredity in Populations and in Pure Lines', in

Selected Readings in Biology for Natural Sciences. University of Chicago Press, pp. 172–215.

JONES, Greta (1980) *Social Darwinism and English Thought.* London: Harvester.

—— (1986) *Social Hygiene in Twentieth-Century Britain.* London: Croom Helm.

JORAVSKY, D. (1970) *The Lysenko Affair.* Cambridge, Mass.: Harvard University Press.

JORDANOVA, L. (1984) *Lamarck.* Oxford University Press.

JUDSON, H. F. (1979) *The Eighth Day of Creation: Makers of the Revolution in Biology.* London: Jonathan Cape.

KALMUS, H. (1983) 'The Scholastic Origins of Mendel's Concepts', *History of Sci.*, 21: 61–83.

KAMMERER, Paul (1923) 'Breeding Experiments on the Inheritance of Acquired Characters', *Nature*, 111: 637–40.

—— (1924) *The Inheritance of Acquired Characteristics.* New York: Boni and Liveright.

KAYE, Howard L. (1986) *The Social Meaning of Modern Biology: from Social Darwinism to Sociobiology.* New Haven: Yale University Press.

KELLER, Evelyn Fox (1983) *A Feeling for the Organism: the Life and Work of Barbara McClintock.* San Francisco: W. H. Freeman.

KEVLES, Daniel (1980) 'Genetics in the United States and Great Britain, 1890–1930', *Isis*, 71: 441–55.

—— (1985) *In the Name of Eugenics: Genetics and the Uses of Human Heredity.* New York: Alfred A. Knopf.

KIMMELMAN, Barbara (1983) 'The American Breeders' Association: Genetics and Eugenics in an Agricultural Context, 1903–1913', *Social Studies of Sci.*, 13: 163–204.

KNORR-CETINA, K. and MULKAY, M. (eds) (1983) *Science Observed: Perspectives on the Social Study of Science.* London: Sage Publications.

KOESTLER, Arthur (1971) *The Case of the Midwife Toad.* London: Hutchinson.

KÖLREUTER, J. G. (1893) *Vörläufige Nachricht von einigen das Geschlecht der Pflanzen.* Leipzig: Ostwaldt's Klassiker der Exacten Wissenschaften.

KOTTLER, Malcolm (1979) 'Hugo De Vries and the Rediscovery of Mendel's Laws', *Annals of Sci.*, 36: 517–38.

KUHN, Thomas S. (1962) *The Structure of Scientific Revolutions.* University of Chicago Press.

LAMARCK, J. B. P. A. de (1914) *Zoological Philosophy.* London. Reprinted New York: Hafner, 1963.

LECOURT, D. (1977) *Proletarian Science: the Case of Lysenko.* London: NLB Books.

LENOIR, Timothy (1982) *The Strategy of Life: Teleology and Mechanics in Nineteenth-Century German Biology*. Dordrecht: D. Reidel.

LEWONTIN, R. and LEVINS, R. (1976) 'The Problem of Lysenkoism', in Hilary and Steven Rose (eds), *The Radicalization of Science*. London: Macmillan, pp. 32–64.

LIMOGES, Camille (1980) 'A Second Glance at Evolutionary Biology in France', in Mayr and Provine (eds) (1980) pp. 322–28.

LOEB, Jacques (1912) *The Mechanistic Conception of Life*. Reprinted Cambridge, Mass.: Harvard University Press, 1964.

—— (1916) *The Organism as a Whole*. New York: Putnam.

LOVEJOY, A. O. (1959) 'Buffon and the Problem of Species', in Glass *et al.* (eds) (1959) pp. 84–113.

LUDMERER, K. M. (1972) *Genetics and American Society: a Historical Appraisal*. Baltimore: Johns Hopkins University Press.

LURIE, Edward (1960) *Louis Agassiz: a Life in Science*. University of Chicago Press.

MACKENZIE, Donald (1982) *Statistics in Britain, 1865–1930: the Social Construction of Scientific Knowledge*. Edinburgh University Press.

MACKENZIE, D. and BARNES, B. (1979) 'Scientific Judgement: the Bio-metry–Mendelism Controversy', in B. Barnes and S. Shapin (eds) *Natural Order: Historical Studies of Scientific Culture*. Beverly Hills and London: Sage Publications, pp. 191–210.

MACROBERTS, M. H. (1985) 'Was Mendel's Paper on *Pisum* Neglected or Unknown?', *Annals of Sci.*, 42: 339–45.

MAIENSCHEIN, Jane (1981) 'Shifting Assumptions in American Biology: Embryology, 1890–1910', *J. Hist. Biology*, 14: 89–113.

—— (1984) 'What Determines Sex? A Study of Converging Approaches', *Isis*, 75: 457–80.

—— (1986) 'Preformation or New Formation – or neither or both?', in Horder, Witkowski and Wylie (eds) (1986) pp. 73–108.

—— (1987) 'Heredity/Development in the United States circa 1900', *Hist. and Phil. of Life Sci.*, 9: 79–93.

MAUPERTUIS, P. L. M. de (1768) *Oeuvres*. Reprinted Hildesheim: Georg Olms, 1968, 4 vols.

—— (1968) *The Earthly Venus*. New York: Johnson Reprint Corporation.

MAYR, Ernst (1942) *Systematics and the Origin of Species*. New York: Columbia University Press.

—— (1959) 'Where are we?', reprinted in Mayr (1976) pp. 307–28.

—— (1972) 'Lamarck Revisited', *J. Hist. Biology*, 5: 55–94. Reprinted in Mayr (1976) pp. 222–50.

—— (1973) 'The Recent Historiography of Genetics', *J. Hist. Biology*, 6: 125–54.

—— (1976) *Evolution and the Diversity of Life*. Cambridge, Mass.: Harvard University Press.

—— (1982) *The Growth of Biological Thought: Diversity, Evolution and Inheritance*. Cambridge, Mass.: Harvard University Press.

—— (1985) 'Weismann and Evolution', *J. Hist. Biology*, 18: 293–329.

—— (1986) 'Joseph Gottleib Kölreuter's Contributions to Biology', *Osiris*, 2nd ser., 2: 135–76.

MAYR, E. and PROVINE, W. B. (eds) (1980) *The Evolutionary Synthesis: Perspectives on the Unification of Biology*. Cambridge, Mass.: Harvard University Press.

MAZZOLINI, R. G. and ROE, S. A. (1986) *Science against the Unbelievers: the Correspondence of Bonnet and Needham, 1760–1780*. Oxford: Studies on Voltaire and the Eighteenth Century, vol. 243.

MEDVEDEV, Z. (1969) *The Rise and Fall of T. D. Lysenko*. New York: Columbia University Press.

MEIJER, Ono G. (1985) 'Hugo de Vries No Mendelian?', *Annals of Sci.*, 42: 189–232.

MENDEL, Gregor (1865) 'Versuche über Pflanzen-Hybriden', *Verhandlungen des naturforschenden Vereines in Brünn*, 4: 3–47.

MILLHAUSER, Milton (1959) *Just before Darwin: Robert Chambers and 'Vestiges'*. Middletown, Conn.: Wesleyan University Press.

MOORE, James R. (1985) 'Herbert Spencer's Henchman: the Evolution of Protestant Liberals in Late-19th-Century America', in John Durant (ed.) *Darwinism and Divinity*. Oxford: Basil Blackwell, pp. 76–100.

MORGAN, Thomas Hunt (1903) *Evolution and Adaptation*. New York: Macmillan.

—— (1916) *A Critique of the Theory of Evolution*. Princeton University Press.

—— (1919) *The Physical Basis of Heredity*. New York: Lippincott.

—— (1928) *The Theory of the Gene*. New Haven: Yale University Press.

MORGAN, T. H., STURTEVANT, A. H., MULLER, H. J. and BRIDGES, C. B. (1915) *The Mechanism of Mendelian Inheritance*. New York: Henry Holt.

MULKAY, Michael (1979) *Science and the Sociology of Knowledge*. London: Allen and Unwin.

MULLER, H. J. (1935) *Out of the Night: a Biologist's View of the Future*. New York: Vanguard Books.

—— (1949) 'The Darwinian and Modern Conceptions of Natural Selection', *Proc. Am. Phil. Soc.*, 90: 459–70.

NÄGELI, Carl von (1898) *A Mechanico-Physiological Theory of Organic Evolution*. Chicago: Open Court.

NEEDHAM, Joseph (1959) *A History of Embryology*. New York: Abelard-Schumann.

NORTON, Bernard (1973) 'The Biometrical Defence of Darwinism', *J. Hist. Biology*, 6: 283–316.

—— (1975) 'Biology and Philosophy: the Methodological Foundations of Biometry', *J. Hist. Biology*, 8: 85–93.

OLBY, Robert C. (1963) 'Charles Darwin's Manuscript of Pangenesis', *Brit. J. Hist. Sci.*, 1: 251–63.

—— (1974) *The Path to the Double Helix*. London: Macmillan.

—— (1979) 'Mendel no Mendelian?', *History of Sci.*, 17: 53–72.

—— (1985) *Origins of Mendelism*, rev. edn. University of Chicago Press.

—— (1987a) 'William Bateson's Introduction of Mendelism to England: a Reappraisal', *Brit. J. Hist. Sci.*, 20: 399–420.

—— (1987b) 'The Role of British Agriculture and Horticulture in the Establishment of Experimental Genetics', unpublished lecture.

OLBY, R. C. and GAUTRY, P. (1968) 'Eleven References to Mendel before 1900', *Annals of Sci.*, 24: 7–20.

OPPENHEIMER, Jane (1967) *Essays in the History of Embryology*. Cambridge, Mass.: MIT Press.

OREL, Vitezslav (1984) *Mendel*. Oxford University Press.

OREL, V. and MATALOVA, A, (eds) (1983) *Gregor Mendel and the Foundations of Genetics*. Brno: Moravian Museum.

PASTORE, Nicholas (1949) *The Nature–Nurture Controversy*. New York: King's Crown Press.

PAUL, Diane (1984) 'Eugenics and the Left', *J. Hist. Ideas*, 45: 567–90.

—— (1985) 'Textbook Treatments of the Genetics of Intelligence', *Quart. Rev. Biology*, 60: 317–26.

PAUL, Diane, and KIMMELMAN, Barbara A. (1988) 'Mendel in America: Theory and Practice, 1900–1918', in R. Rainger, K. R. Benson and J. Maienschein (eds) *The American Development of Biology*. Philadelphia: University of Pennsylvania Press, pp. 281–310.

PEARSON, Karl (1896) 'Regression, Heredity and Panmixia', *Phil. Trans. Roy. Soc. Lond.*, 197A: 253–318.

—— (1898) 'Mathematical Contributions to the Theory of Evolution: On the Law of Ancestral Heredity', *Proc. Roy. Soc. Lond.*, 57: 386–412.

—— (1901) *National Life from the Standpoint of Science*. London: A. & C. Black.

—— (1914–30) *The Life, Letters and Labours of Francis Galton*, 3 vols. Cambridge University Press.

PEEL, J. D. Y. (1971) *Herbert Spencer; The Evolution of a Sociologist*. London: Heinemann.

PICKENS, D. K. (1968) *Eugenics and the Progressives*. Nashville, Tenn.: Vanderbilt University Press.

PIEGORSCH, W. (1986) 'The Gregor Mendel Controversy', *History of Sci.*, 24: 173–82.

POPPER, Karl (1959) *The Logic of Scientific Discovery*. London: Hutchinson.

PORTUGAL, F. H. and COHEN, J. S. (1977) *A Century of DNA*. Cambridge, Mass.: MIT Press.

PROVINE, William B. (1971) *The Origins of Theoretical Population Genetics*. University of Chicago Press.

—— (1986) *Sewall Wright and Evolutionary Biology*. University of Chicago Press.

PUNNETT, R. C. (1907) *Mendelism*, 2nd edn. Cambridge: Macmillan and Bowes.

RAINGER, Ronald (1981) 'The Continuity of the Morphological Tradition in American Paleontology, 1880–1910', *J. Hist. Biology*, 14: 129–58.

ROBERTS, H. F. (1929) *Plant Hybridization before Mendel*. Princeton: Princeton University Press.

ROBINSON, Gloria (1979) *A Prelude to Genetics: Theories of a Material Substance of Heredity, Darwin to Weismann*. Lawrence, Kansas: Coronado Press.

ROE, Shirley A. (1981) *Matter, Life, and Generation: Eighteenth-Century Embryology and the Haller-Wolff Debate*. Cambridge University Press.

—— (1983) 'John Turberville Needham and the Generation of Living Organisms', *Isis*, 74: 159–84.

—— (1985) 'Voltaire versus Needham: Atheism, Materialism and the Generation of Life', *J. Hist. Ideas*, 46: 65–87.

ROGER, Jacques (1963) *Les Sciences de la vie dans la pensée française du XVIII⁰ siècle*. Paris: Armand Colin.

ROLL-HANSEN, Nils (1978) '*Drosophila* Genetics: a Reductionist Research Program', *J. Hist. Biology*, 11: 159–210.

—— (1983) 'The Death of Spentaneous Generation and the Birth of the Gene: Two Case Studies of Relativism', *Soc. Studies of Sci.*, 13: 481–519.

—— (1985) 'A New Perspective on Lysenko?', *Annals of Sci.*, 42: 261–76.

ROOT-BERNSTEIN, R. (1983) 'Mendel and Methodology', *History of Sci.*, 21: 275–95.

ROSE, S., KAMIN, L. J. and LEWONTIN, R. C. (1984) *Not in Our Genes: Biology, Ideology and Human Nature*. Harmondsworth: Penguin.

RUDWICK, M. J. S. (1985) *The Great Devonian Controversy*. University of Chicago Press.

RUSSELL, E. S. (1916) *Form and Function: a Contribution to the History of Animal Morphology*. London: John Murray.

SANDLER, Iris (1983) 'Pierre Louis Moreau de Maupertuis – a Precursor of Mendel?', *J. Hist. Biology*, 16: 101–36.

SANDLER, I. and SANDLER, L. (1985) 'A Conceptual Ambiguity that Contributed to the Neglect of Mendel's Paper', *Hist. and Phil. of Life Scis.*, 7: 3–70.

SAPP, Jan (1987) *Beyond the Gene: Cytoplasmic Inheritance and the Struggle for Authority in Genetics*. New York: Oxford University Press.

SCHAFFER, Simon (1986) 'Scientific Discoveries and the End of Natural Philosophy', *Social Studies of Sci.*, 16: 387–420.

SEARLE, G. R. (1976) *Eugenics and Politics in Britain, 1900–1914*. Leiden: Noordhoff International.

—— (1979) 'Eugenics and Politics in Britain in the 1930s', *Annals of Sci.*, 36: 159–69.

SHAPIN, Steven (1982) 'History of Science and its Sociological Reconstructions', *History of Sci.*, 20: 157–211.

SHRINE, I. B. and WROBEL, S. (1976) *Thomas Hunt Morgan: Pioneer of Genetics*. Lexington: University of Kentucky Press.

SIMPSON, George Gaylord (1944) *Tempo and Mode in Evolution*. New York: Columbia University Press.

SLOAN, Philip (1986) 'Darwin, Vital Matter and the Transformism of Species', *J. Hist. Biology*, 19: 169–445.

SPENCER, Herbert (1851) *Social Statics*. London: John Chapman.

—— (1864) *Principles of Biology*, 2 vols. London: Williams and Norgate.

—— (1884) *The Man versus the State*. Reprinted Harmondsworth: Penguin Books, 1969.

—— (1887) *The Factors of Organic Evolution*. London: Williams and Norgate.

—— (1893) 'The Inadequacy of Natural Selection', *Contemporary Review*, 43: 153–66 and 439–56.

STEELE, E. J. (1979) *Somatic Selection and Adaptive Evolution*. Toronto: Williams and Wallace International.

STENT, Gunther S. (1970) 'DNA', *Daedalus*, 99: 909–37.

STEPAN, Nancy (1982) *The Idea of Race in Science: Great Britain, 1800–1960*. London: Macmillan.

STERN, Curt, and SHERWOOD, E. R. (1966) *The Origin of Genetics: a Mendel Sourcebook*. San Francisco: W. H. Freeman.

STOCKING, George W., Jr (1962) 'Lamarckianism in American Social Science', *J. Hist. Ideas*, 23: 239–56.

—— (1968) *Race, Culture and Evolution*. New York: Free Press.

STOMPS, T. J. (1954) 'On the Rediscovery of Mendel's Work by Hugo De Vries', *J. Heredity*, 45: 294.

STUBBE, H. (1972) *History of Genetics from Prehistoric Times to the Rediscovery of Mendel's Laws*. Cambridge, Mass.: MIT Press.

STURTEVANT, A. H. (1965) *A History of Genetics*. New York: Harper and Row.

SUTTON, Walter S. (1902) 'The Chromosomes in Heredity', *Biol. Bulletin*, 4: 231–51.

SWINBURNE, R. G. (1965) 'Galton's Law: Formulation and Development', *Annals of Sci.*, 21: 15–31.

TEMKIN, Owsei (1950) 'German Concepts of Ontogeny and Development around 1800', *Bull. Hist. Medicine*, 24: 227–46.

VAN BALEN, Gerrit (1986) 'The Influence of Johannsen's Discoveries on the Constraint Structure of Mendelian Research Programs', *Stud. Hist. and Phil. Sci.*, 17: 175–204.

—— (1987) 'Conceptual Tensions between Theory and Program: the Chromosome Theory and the Mendelian Research Program', *Biol. and Philosophy*, 2: 435–62.

VARTANIAN, A. (1953) *Diderot and Descartes: a Study of Scientific Naturalism in the Enlightenment.* Princeton University Press.

WATSON, J. D. (1968) *The Double Helix.* New York: Atheneum.

WATSON, J. D. and CRICK, F. (1953) 'Molecular Structure of Nucleic Acids: a Structure for Deoxyribose Nucleic Acid', *Nature*, 171: 737–8.

WEISMANN, August (1882) *Studies in the Theory of Descent.* London.

—— (1891–92) *Essays upon Heredity and Kindred Biological Problems*, 2 vols. Oxford University Press.

—— (1893a) *The Germ Plasm: a Theory of Heredity.* London.

—— (1893b) 'The All-Sufficiency of Natural Selection', *Contemporary Review*, 64: 309–38 and 596–610.

—— (1896) *On Germinal Selection.* Chicago: Open Court.

—— (1902) *The Evolution Theory*, 2 vols. London: Edward Arnold.

WEISS, Sheila F. (1987) 'The Race Hygiene Movement in Germany', *Osiris*, 2nd ser., 3: 193–236.

WELDON, W. F. R. (1894–95) 'An Attempt to Measure the Death-Rate due to the Selective Destruction of *Carcinas moenas*', *Proc. Roy. Soc. Lond.*, 57: 360–79.

—— (1896) 'President's Address, Zoology Section', *Report of British Association for the Advancement of Science*, 887–902.

WILKIE, J. S. (1956) 'The Idea of Evolution in the Writings of Buffon', *Annals of Sci.*, 12: 48–62, 212–47 and 255–66.

WILLIER, B. and OPPENHEIMER, J. (eds) (1964) *Foundations of Experimental Embryology.* Englewood Cliffs, NJ: Prentice-Hall.

WILSON, E. B. (1896) *The Cell in Development and Inheritance.* New York: Macmillan.

—— (1925) *The Cell in Development and Heredity*, 3rd edn. New York: Macmillan.

WILSON, E. O. (1975) *Sociobiology: the New Synthesis.* Cambridge, Mass.: Harvard University Press.

—— (1978) *On Human Nature.* Cambridge, Mass.: Harvard University Press.

WRIGHT, Sewall (1931) 'Evolution in Mendelian Populations', *Genetics*, 16: 97–159.

YOUNG, Robert M. (1985) *Darwin's Metaphor: Nature's Place in Victorian Culture.* Cambridge University Press.

YULE, G. U. (1902) 'Mendel's Laws and their Probable Relations to Intraracial Heredity', *New Phytologist*, 1: 193–207 and 222–38.

ZIRKLE, Conway (1946) 'The Early History of the Idea of the Inheritance of Acquired Characteristics and Pangenesis', *Trans. Am. Phil. Soc.*, 35: 91–151.

—— (1949) *Death of a Science in Russia*. Philadelphia: University of Pennsylvania Press.

—— (1951) 'Gregor Mendel and his Precursors', *Isis*, 42: 97–104.

—— (1964) 'Some Oddities in the Delayed Discovery of Mendelism', *J. Heredity*, 55: 65–72.

—— (1968) 'The Role of Liberty Hyde Bailey and Hugo De Vries in the Rediscovery of Mendelism', *J. Hist. Biology*, 1: 205–18.

Index

Vitalism, 26, 27, 41, 42, 80, 145

Wallace, A. R., 62
Ward, L. F., 160
Watson, J. D., 173, 175–6
Weismann, A., 3, 31, 53, 74–5,
  83–8, 91, 110, 114, 136, 178
Weldon, W. F., 70, 119
White-eye character, in
  Drosophila, 133

Wilkins, M., 175
Wilson, E. B., 74, 81, 82, 131–2
Wilson, E. O., 184
Wolff, C. F., 40–41
Wright, S., 141–2, 168, 181

Young, R. M., 18
Yule, G. U., 120, 138